Springer Biographies

The books published in the Springer Biographies tell of the life and work of scholars, innovators, and pioneers in all fields of learning and throughout the ages. Prominent scientists and philosophers will feature, but so too will lesser known personalities whose significant contributions deserve greater recognition and whose remarkable life stories will stir and motivate readers. Authored by historians and other academic writers, the volumes describe and analyse the main achievements of their subjects in a manner accessible to nonspecialists, interweaving these with salient aspects of the protagonists' personal lives. Autobiographies and memoirs also fall into the scope of the series.

More information about this series at http://www.springer.com/series/15612

The books published in the Springer Biographies tell of the life and work of scholars, innovators, and pioneers in all fields of learning and throughout the ages. Prominent scientists and philosophers will feature, but so too will lesser known personalities whose significant contributions deserve greater recognition and whose remarkable life stories will stir and motivate readers. Authored by historians and other academic writers, the volumes describe and analyse the main achievements of their subjects in manner accessible to nonspecialists, interweaving these with salient aspects of the protagonists' personal lives. Autobiographies and memoirs also fall into the scope of the series.

More information about this series at http://www.springer.com/series/13617

Luisa Cifarelli · Raffaella Simili
Editors

Laura Bassi–The World's First Woman Professor in Natural Philosophy

An Iconic Physicist in Enlightenment Italy

Editors
Luisa Cifarelli
Dipartimento di Fisica e Astronomia
Università di Bologna
Bologna, Italy

Raffaella Simili
Professor Emeritus
Università di Bologna
Bologna, Italy

ISSN 2365-0613 ISSN 2365-0621 (electronic)
Springer Biographies
ISBN 978-3-030-53964-1 ISBN 978-3-030-53962-7 (eBook)
https://doi.org/10.1007/978-3-030-53962-7

Cover illustration: Portrait of Laura Bassi. Carlo Vandi, oil on canvas, 18th century, Bologna, Museo di Palazzo Poggi.

This Springer imprint is published by the registered company Springer Nature Switzerland AG
The registered company address is: Gewerbestrasse 11, 6330 Cham, Switzerland

Preface

Laura Maria Caterina Bassi Veratti, known as Laura Bassi, was born in Bologna in 1711. In 1732, at the age of 21, she became the first female member of the influential *Accademia dell'Istituto delle Scienze* in Bologna and was the first woman in the world to be appointed to a University chair to teach "universal philosophy" (1732) and then "experimental physics" (1776).

On the occasion of the tercentenary of her birth, the University of Bologna organised an exhibition, a number of lectures and a conference, and the *Accademia* itself established a "*Forum Laura Bassi*" which since then, every year, focuses on themes of topical interest. On the same occasion, the *Società Italiana di Fisica* (SIF) and the *Società Italiana di Storia della Scienza* (SISS) have taken the initiative—incidentally, being both led then by female professors of the University of Bologna and at the same time members of the *Accademia*—to publish a volume on Laura Bassi to pay a tribute to her in a way that would leave a lasting memory for the future and could mark a starting point to promote further analyses. And so the idea of the volume "LAURA BASSI: EMBLEM AND PRIMACY IN SETTECENTO SCIENCE" was born. This volume, appeared in 2012, was a collection of writings by various experts which aimed to illustrate the personality and work of Laura Bassi, as well as the characters who most influenced her education and her research activity, and, at the same time, to underline the main features of the physics of the eighteenth century in the frame of which this scientist operated. To make it easier to read and to foster the widest possible circulation, the volume consisted in an Italian version at the beginning and an English translation at the end. A plentiful selection of attractive and telling pictures of the epoch, pertaining to Laura Bassi, the scientific context of the time and particularly the instrumentation referred to in the various essays, provided a lively and colourful interval in the passage from one language to the other.

A few years later, in 2019, a "Laura Bassi Medal", intended for its most meritorious PhD graduates, has been established by the University of Bologna (among other similar medals named after Petrarch, Copernicus, Accursius and Marconi). In 2019, the new icebreaker ship of the *Istituto Nazionale di Oceanografia e Geofisica Sperimentale* (OGS) for scientific research in the Antarctic has been named "Laura

Bassi". And in 2020, a special documentary dedicated to Laura Bassi has been produced,[1] which will be broadcast on the major Italian national television channel in the framework of the very popular and renowned show "*RAI Storia*".

For all the above, the SIF, in collaboration with Springer, has decided to publish the present volume in the Springer Biographies series, as a new revised and updated edition of the previous one, in order to revamp and widen the knowledge of this extraordinary woman.

The book opens with an essay by Walter Tega who examines in an innovative way the distinguished figure of Luigi Ferdinando Marsili, scientist, oceanographer, soldier, member of the Royal Society and well known in Europe precisely for his exceptional gifts as a naturalist, the guiding spirit behind the creation of the *Istituto delle Scienze* which was founded in 1711 when its *Costituzioni* were approved on 12 December.

The project devised by Marsili was ambitious and original, aiming to house the entire encyclopedia of modern scientific knowledge in the rooms of an ancient senatorial residence in Bologna. A series of laboratories, galleries, workshops, laid out over the two floors of Palazzo Poggi (to which must be added the Specola (Observatory), designed in 1711 and only completed in 1726, and the library, built according to a design by Carlo Francesco Dotti, created according to the wishes of Pope Prospero Lambertini in the middle of the century), made the *Istituto*, from its very beginning, not just a place for discussion similar to that offered by the numerous scientific academies active within the *République des Lettres*, but also a venue where the sciences could be cultivated experimentally, an unquestionably rarer thing.

It was also Lambertini, first as archbishop of Bologna and then as Pope Benedict XIV, who supported Marsili's enterprise and relaunched it in the Europe of the Age of Reason. The guidelines for reform suggested for the Istituto and to its Accademia, the considerable increase in its resources, to the library, to the collections, to the equipment of the laboratories, the bold expansion of the network of foreign correspondents, the need felt by the Roman Curia to enhance scientific knowledge, at an institutional level too, at the most acute phase of the harsh anti-religious controversy opened up by the Enlightenment, constituted the framework within which men of science like Manfredi, Zanotti, Molinelli, Beccari, Galeazzi and their pupils Laura Bassi, Leopoldo Caldani and Luigi Galvani captured an eminent position in the scientific culture of the eighteenth century.

This experimental activity and the empirical observational method formed the inheritance that Marsili left to the *Istituto* and that we encounter in Laura Bassi enriched with a new method to conceive the teaching of physics, typical of the eighteenth century.

As said, Laura Bassi was born in Bologna in 1711. Her father Giuseppe was a lawyer from Scandiano, and her first teacher was a cousin, father Lorenzo Stegani, who taught her grammar, Latin, French and arithmetic; she then studied natural

[1]The documentary is sponsored by the *Museo Storico della Fisica e Centro Studi e Ricerche "Enrico Fermi"*, the *Società Italiana di Fisica* and the University of Bologna.

philosophy with a doctor, Gaetano Tacconi, who was lecturer of medicine at the University and a member of the *Accademia delle Scienze*.

From 1731, the archbishop of Bologna, Cardinal Lambertini, was one of her supporters: since then, he followed her progress in her studies, becoming her strongest and most authoritative patron.

On 20 March 1732, she was named honorary member of the *Accademia delle Scienze*; on 17 April of the same year, she presented her thesis in public, as was the custom at the time, and on 12 May, she was awarded a degree in philosophy. On 29 October, she obtained ex officio from the Senate a post to teach universal philosophy, also due to pressure from Lambertini, regularly salaried by the University.

Her public role became that of woman prodigy, equipped with an intelligence that was extraordinary in that she was a woman, a wonder to be shown off; her fame also reflected on her city, Bologna. Laura showed herself in the most prestigious cultural venues, such as the *Archiginnasio* and the *Accademia delle Scienze*, always on occasions of events accompanied by splendour, magnificence and by immense social and political participation.

By her marriage in 1738 to Giuseppe Veratti, a doctor and expert in physics, her authoritative colleague at the University and in the Academy, she had eight children, five of whom survived.

In 1745, she became a member of the Benedictine Academy: it was Bassi herself who persuaded the pope to create a supernumerary twenty-fifth place designed for her, in addition to that body's roll of twenty-four members. With this appointment, although again she was not guaranteed parity in terms of the rights enjoyed by the other members of the Academy, she achieved the highest recognition for her scientific activity.

From 1749, she organised, with her husband's help, a private school of experimental physics in her home, with courses that were imbued with Newton's method.

In the Verattis' laboratory, where the couple carried out their most original investigations including the medical use of electricity, experiments were also conducted by the doctor of medicine from Bologna Leopoldo Caldani and the naturalist Felice Fontana from Trento, disciples of the new theories on muscular irritability put forward by the Swiss naturalist Albrecht von Haller. The laboratory had the necessary instruments available for the preparation of Laura's reports to the Academy and for the debate she engaged in, by correspondence, with scientists such as Antoine Nollet, Giovanni Battista Beccaria, Felice Fontana and Lazzaro Spallanzani.

Her scientific interests ranged from optics to analytical mechanics, to hydrometry, to electrology, to pneumatic physics and chemistry. She began to study Newtonian physics and infinitesimal calculus with Jacopo Bartolomeo Beccari and Gabriele Manfredi: research that would allow her both to offer her own students innovative courses in experimental physics and also to become one of the protagonists in the acceptance and circulation of Newton's theories in Italy.

From the sixties, her overwhelming curiosity was in the field of electrical phenomena, as can be inferred from the titles of some of her communications in this field of study—now unfortunately lost—and the exchange of letters with some of the greatest specialists in this sphere. Only four of Laura Bassi's memoirs presented to the *Accademia delle Scienze* in Bologna, in which she dealt with questions of hydrometry, mechanics and pneumatic physics, have come down to us. In 1766, she was appointed to teach experimental physics at the *Collegio Montalto*. In 1776, she was given the post of professor of experimental physics at the *Istituto delle Scienze*, with her husband as her "substitute". It was he who would replace her there on her death in 1778.

Coming back to our volume, with regard to Newton's theories we turn to the lucid contribution by Niccolò Guicciardini who, after an acute analysis of the main aspects and problems of these theories, which Newton himself left as his legacy to the scientific community, points out how Laura Bassi's physics, while appearing on the one hand to be influenced by Newton's *Opticks* and *Queries*, reveals on the other hand a strong fascination with mathematical physics inspired by the *Principia*. The *Queries*, as is well known, are "open questions" that Newton left as if he wanted to suggest lines of research still to be completed. In particular, it should be noted that the Queries that Newton had stressed with regard to chemical and electrical phenomena showed an attention for the phenomena of perception and volition, that is to say for vital phenomena, an attention shared by the Bassi-Veratti couple. In conclusion, it can be claimed that Laura Bassi and the Italian eighteenth-century Newtonian scientists fully accepted the open nature of Newton's legacy, carrying out wide-ranging research to this aim.

The accurate and well-documented contribution by Sofia Talas specifically deals with *Settecento* physics and elucidates some important choices of Laura Bassi. Sofia Talas, on the basis of a long excursus on the history of the importance of instruments in physics since Galileo Galilei, brings to light, among other things, the extraordinary success enjoyed from this point of view by the introduction of electricity, in laboratories, in academic circles and even in salons. It was in this context that Laura Bassi also began to work on electricity, so much so that the already mentioned Antoine Nollet, a well-known figure in this field, met her at Bologna during his journey to Italy in 1749, remaining in correspondence with her. Nollet dedicated to her one of his *Lettres sur l'électricité* published between 1753 and 1767, in which he presented his most recent experiments and his own theories on electricity.

Paula Findlen's essay, with the significant title Amongst Men, provides the central theme of the book since it tells in detail the intricate and difficult story of "*Signora Laura*", a scientist certainly but also the only woman amongst men, University teachers and members of the Academy.

One of these men, the medical doctor Giovanni Bianchi from Rimini, a remarkable figure in Italian eighteenth-century culture, on several occasions showed a special sensibility towards female involvement in the scientific community, as Miriam Focaccia. accurately explains in her article. Focaccia stresses that personal relationships and correspondence with several female exponents of the culture of the time were of particular interest, including the mathematician Maria Gaetana Agnesi from Milan and, above all, the two women from Bologna Anna Morandi Manzolini, the anatomist, and, of course, Laura Bassi.

Finally, the laboratory of the Bassi-Veratti couple illustrated by Marta Cavazza is linked to Findlen's essay, showing once again Bassi's innovative teaching and the experimental spirit of the institute which remained intact in her.

Laura the scientist was an emblem, an iconic figure: then, as a woman, a primacy and a role model. Between women, between female scientists, we understand each other; for that very reason we wanted to introduce to a wider public in Italy and abroad, not just symbolically, the extraordinary intellectual legacy that she left us from that distant and marvellous period that was the Settecento.

University of Bologna
January 2020

Luisa Cifarelli
Raffaella Simili

Contents

1 **Luigi Ferdinando Marsili's Science Institute and Its Academy** 1
Walter Tega

2 **Newton's Legacy: An Open Field of Research** 35
Niccolò Guicciardini

3 **Physics in the Eighteenth Century: New Lectures, Entertainment and Wonder** 49
Sofia Talas

4 **Always Among Men: Laura Bassi at the Bologna Academy of Sciences (1732–78)** 69
Paula Findlen

5 **Giovanni Bianchi: A Sensitive Promoter of the World of Women** 95
Miriam Focaccia

6 **The Bassi-Veratti Home Laboratory** 115
Marta Cavazza

Credits 143

Index of Names 145

Chapter 1
Luigi Ferdinando Marsili's Science Institute and Its Academy

Walter Tega

1.1 Marsili in His City

Marsili spent just over fifteen years of his adult life in Bologna. He left his city in 1680 to return and settle there, albeit with some important breaks, from 1708 to 1726. It must be said that they were intense years, both in the fields of politics and diplomacy and also in the scientific field. In essence, considering the general's life style, the specificity and the methods of the routes he followed, we can say that his stay in Bologna very soon assumed the same character as his journeys which were rooted in science. Indeed it was his longest and most challenging journey. He was not facing the Ottoman enemy, nor cities or sites to fortify or storm, nor frontiers to define, nor even lands and seas to observe and study. On the one hand there were the quicksands of the academic establishment, on the other the pitfalls of the "mixed government" and the intervention and mediation of the Roman curia (there was actually more than a little resemblance to the imperial court). Also there was his firm determination to give true shape to a scientific project that was really original in Europe and that he had sought for a long time. To all this was added the desire which lasted well into the 1720s to bring order to all his vast quantity of jottings and to the writings accumulated over his military and scientific campaigns. So Luigi Ferdinando Marsili returned to settle in Bologna after twenty five years of political, military and diplomatic experience in various parts of Europe (it was not yet possible to speak of his scientific experience for the simple reason that the two great works that made Marsili famous as a naturalist were published from the mid-'20s), studded with success and acknowledgements but also accompanied by setbacks, defeats and dramatic events that deeply marked his life. His relationship with his city and his family of origin was never short of misunderstandings and conflicts but nevertheless

W. Tega (✉)
Emeritus, Bologna University, via Zamboni 38, Bologna, Italy
e-mail: walter.tega@unibo.it

L. Cifarelli and R. Simili (eds.), *Laura Bassi–The World's First Woman Professor in Natural Philosophy*, Springer Biographies,
https://doi.org/10.1007/978-3-030-53962-7_1

the ties of memories and attachments linking him to them both were never completely severed. He cherished a sense of gratitude and felt a closeness towards the city that accompanied him in all the events of his life; but it must immediately be said that his constantly evoked sense of gratitude had as an essential point of reference his youthful studies under the guidance of excellent and never forgotten teachers such as the "divine" Malpighi, Geminiano Montanari and their pupil Trionfetti. It was their example that had awoken in him the demon of research, directing him clearly towards a tempered Galileism and towards the adoption of experimental procedures to which the destiny and success of the new science were linked. The closeness was represented on the one hand by his feeling part of a long historical process that had consecrated his city as a centre of excellence of knowledge, as a teacher of civilisation for the whole of Europe, on the other by having kept close ties with his old fellow students who, in the meantime, had become outstanding teachers at the University. They had had the merit of keeping a scientific perspective and practice alive in the city for which there was no room in official quarters but which had developed in the confines of a fellowship which Marsili always looked on with care and solicitude, even from afar. These were the members of the Accademia degli Inquieti (Academy of the Restless) who would play a crucial role in the initiative that the general was preparing to launch after returning to the city.

In 1711 Marsili was bringing to a conclusion a laborious negotiation which had begun over a year earlier with the Holy See and the city Senate. From their economic and institutional commitment, and from the gift of his precious collection of books, scientific instruments and natural and archaeological finds, the fruit of the passion of a man of science rather than a soldier, a scientific institution was born which had to maintain official relations with the University, with the undeclared aim of introducing the new scientific attitude that had matured elsewhere in Europe but that in his city had not put down roots, except for rare exceptions and fortunate moments. It would take a further three years of gruelling political and diplomatic debate, accompanied by an equally long gestation of his detailed scientific project, before the Istituto delle Scienze e delle Arti (Institute of Sciences and Arts) would take form and would begin to function. It was 1714 when, in the prestigious building that had belonged to cardinal Poggi and had been made available by the city's Senate, Ercole Corazzi gave the inaugural oration and the Inquieti were allotted the task of defining the structure and promoting the activities of the succession of workshops "made for eyes and ears" and open to public use, where the demonstrations, lessons or lectures analogous to those taking place in the Archiginnasio, the seat of the University, were not allowed to be substituted.[1]

This journey through his city did not spare Marsili any of the obstructive difficulties connected with policy procedures, scientific organisation and what we might euphemistically call diplomatic relations.

So Marsili wanted to tackle the crisis of the city, putting forward a diagnosis and indicating remedies, and to this aim he prepared a memorandum for the Senate and for the Assunteria di Studio (the administration of the university) regarding the

[1] Angelini (1993).

comparison between the university of Bologna and those "beyond the mountains". As a citizen who had cultivated "sentiments of gratitude" towards his country he felt called upon by "obligations of nature and honour" to remind the Bolognese that their forefathers "laboured to restore those sciences that so many barbarous nations had trampled" but, at the same time, to emphasize that those same forefathers did not only think of renewing those sciences for the use of their own citizens but also to extend the benefit and fruits of such a vast institute to the advantage of the farthest nations, which competed to rush to this city of ours, like children to their mother's breast, to feed on this scholarship that the city strived to be able to communicate, and to know how to communicate, to the whole of Europe. Thus Marsili stressed not only the capacity displayed by the city when it conceived this model but also the precise conviction that this model took on its true and real significance by the simple fact that others too, in other countries, could benefit from its civilizing characteristics. And indeed this was his city's destiny, this was the only route which could lead it out of its crisis and reconquer the eminent role that it had irretrievably lost: "a single city such as this one, with a small territory, could in no other way become famous except by means of literature, which is acquired without the aid of armies or vast domains, but only by application and by exercising one's wits".

In this booklet, provocatively titled *Parallelo dello stato moderno della Università di Bologna con l'altre di là de' monti (Parallel between the modern state of the University of Bologna with the others beyond the mountains)* (1709) Marsili intended to point out, without beating about the bush, both the lag built up by the ancient University compared to European scientific institutions and also to indicate the routes that should be followed to put it forward to international attention again by reconsidering its scientific and organisational options.

According to Marsili it was necessary to equip the ancient University and put it in a position to make the best use possible of the financial means, which were far from being insufficient, assigned to it by the city and the papal government, to let it participate in the new sciences, to adapt it to the new requirements of research and training so as to return it to its European dimension.

The list of necessary interventions was rather long: amongst the questions to be dealt with, as a priority, there were some that were particularly challenging. The first concerned the incentives that a public power interested in the growth of knowledge could not fail to provide in order to spur on the initiative of those who "are capable of making new discoveries in the sciences". The second concerned the working conditions of the teachers who should be reduced in number and paid more, taking care to distinguish between them and, in distinguishing, to favour the pay of professors of Mathematics, of the Natural Sciences, of Languages, since they are all sciences that do not lead to the day to day usefulness afforded by lawyers and doctors of medicine".[2] The third question concerned the innovations and modifications to be introduced into the programme of studies and the direction to be given to research. Marsili was aiming his sights forcefully at mathematics which between the 16th and

[2]L. F. Marsili, Parallelo dello stato moderno della Università di Bologna con l'altre di là de' monti, in E. Bortolotti (1930).

17th centuries had earned Bologna absolute prestige and that needed to be revitalized by enhancing algebra and the applied aspects of mechanics, of military art, of planimetry, of geography and, naturally, of astronomy which, after Giandomenico Cassini's departure for Paris, seemed to be heading into an unstoppable decline. Although apparently less affected by the crisis of the late 17th century, the study of medicine also had to be revived and to do that it would be necessary to tread in "the footsteps of the Divine Malpighi". In other words to develop its overwhelming proximity with the fields of physics "the true knowledge of the study of nature", of the "practice and institutions of medicaments" and, above all, of chemistry, a discipline still missing in university curricula but essential, and therefore to be introduced as soon as possible amongst the Chairs of the University according to Marsili "since it is impossible to be a good experimental physicist in the modern usage, without having a perfect knowledge of chemistry".[3] But precisely because Marsili the politician knew that scientific life and civil life were intertwined, even the sector of lay history—also excluded from university teaching—had to be fostered, supported by "a perfect chronology" and entrusted to a teacher (really still difficult to find today) familiar with mathematics, an expert in the Greek language and not unaware of astronomy. The general who had travelled the Danube region, the diplomat who had contributed to the signing of the peace of Karlowitz, the geographer who had studied in depth the cartography of the Arabs as well as of Europe, had not failed to notice the urgent need to establish teaching courses in the oriental languages from which a positive outcome could be expected not only for sciences and for arts but also for politics and religion.

When Marsili put forward this proposal he cannot have held great expectations of success because he also had before him the vivid memory of the setback suffered, and the price paid, ten years earlier by the proposal to reform the University, much more limited in scope both scientifically and in terms of organisation, put forward from a much more authoritative position by his elder brother Anton Felice who held the office of High Chancellor of the University and Chairman of the degree sessions and who, at the end of his unfortunate struggle, had to abandon the city (1701) after having obtained with some difficulty the bishopric of Perugia. Also Luigi Ferdinando foresaw that the opposition of the academic establishment itself and of the more conservative part of the Senate, would be so forceful that any proposal for change would come to nought. And besides his experience of command made him conscious that an alignment of the city's University with those that "were beyond the mountains" would be virtually impossible for many reasons, economic and structural as well as of political favouritism. His diagnosis, however was severe and the cure harsh. His was certainly an authoritative provocation designed to arouse concerns, annoyance, enmity but also doubts and approval. Marsili's text was discussed even in Rome, but it does not appear to have been discussed in any of the city's official bodies. Very soon interest in the *Parallelo* waned, yet even so it left its mark in the city's more open minded circles and, above all, in the Roman curia which did not look favourably on the stagnation reigning in Bologna. But this may actually

[3] Bortolotti (1930).

have been the result the general had been aiming for. Besides that he knew equally well—and here lay perhaps the real political force of his proposal, which was not as naive as the Assunteria di Studio may have thought—that the decision to donate his library, his scientific instruments, his natural science collections and archaeological finds to the city to assist the renewal of research would place the Senate in grave difficulty[4] These are the reasons that lead me to insist on the fact that the main aim pursued by Marsili's intervention was not so much the reform of the University but his fallback position compared to his opening gambit, aiming at the creation in Bologna of a completely new institution which might even be able to scourge the ancient University. The intent was to open up the path to the establishment of an Istituto delle Scienze e delle Arti (Institute of Science and Arts) entirely dedicated to experimental research and capable within a short while of taking its place amongst the circles that were delineating the advancement and affirmation of the new sciences in Europe. Indeed the "edifice" spoken of by Marsili, the "Atlantis achieved" to which Fontenelle referred or "the work of a king", as Boerhaave had called it, needed time and it could not be expected to spring, like Pallas Athena, already armed from the brain of Zeus.

Marsili believed that his working hypothesis might have some chance of success. He had thought about it for a long time, during his military campaigns and his deliberations together with the Inquieti, but all that time the reform of the university had never been officially mentioned. And indeed, once the danger of a reform had been set aside, the academic establishment could have remained in their quiet state but the city's Senate would have to do something faced with such an attractive and pressing proposal as the general's substantial and generous donation which was now being talked about in the city with great interest and which was warmly appreciated in Rome.

It is in fact legitimate to believe that within this complicated strategy Marsili was also counting on the benevolence and support of the pope—for whom he had acted as general and military consultant—if it is true, and it is, that the complicated dealings leading to the birth of the new institution were brought to a conclusion in a remarkably short time. In February 1711 the complicated matter of the donation was concluded along with the identification and purchase of the building to house the new Istituto. Marsili's strategy had arrived at happy outcome, the new sciences now had a home. Now all that remained to be done was to arrange them adequately [5].

[4]The value of the range and importance of the cultural heritage that Marsili was preparing to give to the city was noted favourably even by an observer who had always been critical of the general such as the commentator Ghiselli (Biblioteca Universitaria Bolognese, hereafter BUB, ms. Ghiselli 75, ff. 30–38); see also Stoye (2012).

[5]Much has been written about the controversial but nevertheless long relationship between the Istituto delle Scienze and the University, see especially Tega (1984); A. Angelini (1993).

1.2 From the Danube to the Science Institute: The Scientific and Institutional Journeys

It is not easy to follow the phases that lead to the definition of the Istituto's structure, even the deliberations on its scientific duties and its rules of behaviour had engaged the general and the whole group of the Inquieti for many years. Probably Marsili began in practice, immediately after bringing the *Prodromus danubialis* to a conclusion, when he decided in 1702 to entrust to a text, destined to remain a manuscript and given the title *Note per l'Accademia* (Notes for the Academy), his proposal for a public institution to be established in his city, not aligned with the state and the procedures of the University, and devoted exclusively to the study and practice of Astronomy, Mathematics, Experimental Physics, Chemistry; governed by a selected number of experienced men of science; equipped with laboratories and instruments and a printshop to ensure the circulation throughout the "Republic of Letters" of the observations, reports and experiments carried out, entrusted to a real scientific diary of the institution.[6]

I do not believe that this text, written in the Forest of Hagenau, can be separated from the parallel research and the acquisition of books and instruments that Marsili had accumulated in those years of study and campaigning and that he intended for Bologna. It was from here that his friend Trionfetti sent news about the construction of the observatory, a task Marsili had shouldered since the departure of his brother Anton Felice for Perugia. The arrival of Marsili's instruments coincided with the process of reviving the scientific activity of the Inquieti who found in Morgagni the inspirer of new rules, drawn up in 1704 and imbued with the rules that guided the scientific activity of the Académie des Sciences in Paris, and a new home in palazzo Marsili.

It must be repeated that without the support of these young men of science who divided their time between the University and their experimental activity carried out in makeshift laboratories, with equipment provided for with great generosity by Marsili, it would have been difficult to launch the initiatives that would lead to the creation of the Istituto. It was in fact the leading lights of this brotherhood who supervised the outfitting of the laboratories and the arrangement of the books, of the scientific instruments, of the natural history collections that would constitute the essential core of the future institution. When indeed Marsili came to sign the agreement with the Senate and with the Roman curia about this new public institution he immediately agreed that the Accademia degli Inquieti should become a constituent part of the enterprise, taking on the more solemn name of the Accademia delle Scienze dell'Istituto (Academy of Sciences of the Institute). Again it would be Marsili, together with these young men of science, who would draw up the Istituto's constitution and who would adopt the sign that Guidalotti had painted for the Inquieti,

[6] Marsili, *Punti per l'Accademia*, in *Libro di più miei pensieri che mi venivano alla giornata*, 1702 (BUB, Mass, Marsili, 83B, cc. 79–83).

who were now changing home and name, as a symbol of continuity of tradition with the new.[7]

It would nevertheless take three years to establish the Istituto in the new building which clearly showed its characteristics as an alternative compared to the University, with which however it would be necessary to maintain a collaborative relationship which would intensify in the years of the "principality" of pope Benedict XIV. At the moment of the official inauguration the scientific route laid down in the rooms of the ancient Palazzo Poggi first took shape; it was not an easy undertaking, also because it was necessary to house a sequence of laboratories and required material for research that represented knowledge and disciplines following criteria that were totally different not only compared to the organisation of the University but also of all the foreign institutes which had been adopted as examples. So Marsili's "edifice" was something more, from the point of view of instrumentation, both with regard to the model offered by the Académie des Sciences, and the model of the more accredited Royal Society. As well as the fittings of the Accademia Clementina (room for nude drawing, moulds of ancient statues) which, by the wish of Pope Clement XI and to the great pleasure of Marsili who had become its patron, was aggregated with the Istituto, the succession of rooms provided for the Library, the Antiquities room, the rooms for Physics, Chemistry, Architecture and Military Art, Natural History, the Workshops for lathes and the other tools necessary for the laboratories to function and for demonstrations and last, but obviously not least, the Specola (Astronomic Observatory) which, according to the agreement reached with the Senate, would have to be built from scratch in the building where one was lacking. This operation already began in 1712 but was only concluded in '26.

The layout of the building and its rooms proposed a strategy for scientific knowledge that was completely unknown to the University; to give birth to an institution that would allow disciplines to be truly in contact with each other, to encourage unheard levels of cooperation between scientists so as not only to facilitate exchanges and relationships but to give new dimensions to research that on the one hand would postulate a real relationship between theory and practice and on the other would inaugurate that work at the frontier that constituted the true and fertile nature of the new *modus operandi* of the scientist. The design was that of a working scientific encyclopedia, capable of being placed at the foundation of the full development of the dynamics of knowledge, of maintaining relationships and of understanding affinities and kinships between the new segments of knowledge and action which would eventually add new tesserae to the overall picture of the mosaic of knowledge, going so far as to modify its appearance incessantly.

I believe that it is only within this architectural arrangement of knowledge and action that it could have been possible to bring faithfully to light the intention with which Marsili set out to seek and select the items that were to constitute his donations. He was not driven by the passion of the professional collector, he was not a refined seeker after marvellous and extraordinary things to house in his wundercamera, much less was he an acquisitive and curious soldier who took advantage of war and

[7]On the emblem of the Inquieti see A. Angelini (1993), pp. 188–192.

the opportunities that it offered to appropriate those things of value saved from the destruction resulting from sieges and conquest. Marsili was an expert researcher and a wise purchaser of books and instruments for which he had in mind tasks, functions and goals. Indeed these are the keystones to reading his working scientific encyclopaedia, and it is this that marks out the new type of activity which, apart from the often sterile polemics that had raged in the recent past (ancient *versus* modern; book-based culture *versus* practical ability; attention to sciences *versus* interest in the mechanical arts) had chosen the path of integration between book, laboratory, observation in the field, scientific reflection and, finally, its practical and instrumental application.

Books represented tradition, which always has something to teach us if adopted outside any dogmatic principle of authority; books bore witness to journeys of exploration, of direct observation of nature; books as scientific treatises capable of showing the level of complexity, abstraction and generalisation reached by individual disciplines; books as auctoritas capable of descending from the pedestal of its doctrinal systematics to be broken up into inserts, leaflets that evolve into diaries, into periodic publications aimed at ensuring the proliferation and circulation of the scientific results achieved by Academies or public and private scientific Societies scattered throughout the Republic of Letters.

And Marsili, apart from any indulgence towards book culture, showed that he really believed in the function of this tool so that he dedicated reflections of considerable importance and constant research to organising a library which did not claim universality but which would nevertheless possess two special characteristics. It would gather together the essential and the best of all that is knowable, and it would be comprehensive and up to date with regard to the disciplines the Istituto revolved around. Essentially it was a truly encyclopedic library as far as the scientific disciplines were concerned, encyclopedic in that it was designed to act as a necessary support for the specialist so that he could understand that his specialism would never be able to acquire a connotation that was both rigorous and developing without the aid and contribution of many other disciplines.

At this point it would be appropriate to mention two further observations which confirm, if only indirectly, Marsili's strategy with regard to the establishment of the Istituto as a new and autonomous entity. Firstly: Marsili already began to think about a library suitable for a truly innovative research institute in 1703.[8] Secondly: the route chosen to define the plan of this library adopted an operational procedure that was made to measure for an institution that had to be created *ex novo* according to rigorous scientific criteria and that was thus totally unsuited to an antiquated, indolent and basically unreformable University such as the University of Bologna. To provide greater substance to his plan Marsili asked Trionfetti, Manfredi and Stancari to help him to compile a catalogue of "desiderata" to send to the major booksellers and antiquarians in various countries in order to then proceed with purchases. When his first donation was signed in 1712, the library had already been established and its

[8] For the creation and composition of Marsili's "library" see the deeds of gift of 1711 and 1727 and above all Bianconi (1894).

structure would remain essentially identical even after the second donation, in '27, which brought to Bologna the books purchased with the sum made available by the Dutch editors for the publication of *Danubius*.

Certainly the idea of compiling a catalogue for the establishment of a library was not new, Marsili handled Hartley's catalogue skilfully[9], but he was also familiar with the catalogue of Gabriel Naudé, amongst others, and in Vienna he must also have heard of the project that Leibniz had created more recently for the Emperor's library[10]; but the catalogue of what would become the library of the Istituto is of remarkable interest not just for its systematic nature or its size, but for its selectivity both relating to the overall picture of what can be known and with regard to individual disciplines.

If indeed the sections dedicated to Politics, lay History and religious History could count on a limited but highly select number of works, the same would not apply to the sectors that accounted for the Istituto's scientific lifeblood.[11] The first core of the library was indeed made up of volumes designed to constitute "a perfect series for the study of nature and that covered Botany, Zoology, Mineralogy, Hydrography, Agriculture"; immediately after these came Mathematics, in its widest sense of the theoretical and applicational articulations that this discipline had evolved over the past century. And so, alongside Geometry, Astronomy and Trigonometry, were set Mechanics, Geography, Navigation, civil and military Architecture, Ballistics, Hydrology, Optics, Stereometry, Gnomonics. To natural history and mathematics were added the highly relevant section on ancient erudition and archaeology which had, amongst other tasks, that of putting nature and artifice in sequence. But within this extraordinary architecture of knowledge the lexicons could not be omitted both of the western languages and, in accordance with the appeals from his personal experience summed up clearly in his *Parallelo*, the Oriental Languages. But, as we were saying, the range of books contained in Marsili's library, gave ample room not only to the projects and to the catalogues of the great libraries, but recognised the great prominence acquired in terms of the circulation and communication of new scientific developments, journals, leaflets, the scientific diaries of academies and of public and private research institutions.

But there were three more sections of the "library" which could not be allowed to lack the characteristics of completeness and rigour that Marsili demanded; these were Physics, which had made extraordinary progress thanks to Huygens, Newton, Musschenbroek; Medicine—which in Bologna was also at home at the University—where anatomy, surgery and pathology had produced exceptional protagonists who, however, had to move to other universities; Chemistry which Newton and Boerhaave had placed in an eminent position in their research and that struck Marsili as the discipline of the future.

But our reader will certainly not be surprised to learn that a significant and by no means secondary sector of this fascinating library was dedicated to travel literature,

---

[9] Hartley (1699).

[10] Naudé (1627), Leibniz (1689).

[11] R. Gherardi (1975).

or rather the accounts and diaries of explorations, of naturalist observations carried out in the field. Nor could the collection of Arabic. Hebrew and Greek manuscripts, of which count Marsili had had drawn up a complete catalogue, be overlooked.[12]

It is certainly not by chance that for Marsili the library ought to occupy the first rooms of the Istituto[13].

The same could be said for the instruments that were acquired with scientific skill and with his mind turned to an idea. The endowment of instruments was the second keystone to understanding the design and aims of the project, instruments in fact were to interact both at the level of book culture, and at the level of experiments and observations that represented the incipit of the experimental path. They were the pivot around which activity revolved in the laboratories that distinguished this place of research from the lecture halls of the University, countering the laziness of those men of science who thought that they could examine nature while remaining shut up in their library or their own laboratory. On this subject too, in Vienna when Marsili together with Einmart devised an example of the "pneumatical engine" according to Boyle's indications or selected the instruments to equip the observatory that the Inquieti were setting up in his building, Marsili was not really thinking of a reform of the University, but of something fundamentally different, untried and exceptional for his city. So the air pump, the burners and stills, the quadrants and telescopes, the models regarding the science of war, the natural history and archaeology collections, the lathes all constituted the necessary trait d'union between the work carried out in the laboratory and activity in the field, in the sense that using them could contribute to identifying a line of research, but it was to them that one had to return when it became necessary to move from the world of conjecture and approximation to the world of identifying laws.[14]

So this was the strategy that he adopted as the basis for the division of the building into rooms that became the actual locations for this complicated process founded on Mathematics, Astronomy, on Natural History and, above all, on Chemistry which, in his opinion (and in this he could take comfort from men like Boerhaave and Newton), constituted the most advanced moment in scientific research of the age. For this reason he never tired of postulating a better way of functioning for this area of study that his friends found hard to understand and his insistence on establishing a specific chair that was the first of its kind in Italy. But the "edifice" had other important instruments available for its growth and for its affirmation beyond the confines of the city. The first consisted of the function that the Academy had to take on as a seat of communications and of exchange with regard to the progress made in research and the results achieved in the laboratories in via Zamboni (where the Academy was located) and those from the whole Republic of Letters; a body made up of real members and of correspondents, which carried out the function of both resolving doctrines and theories and of trying to piece together, in moments of temporary convergence, if not actually unity, the various pieces of research that passed through the private and public

[12]Talman (1702).

[13]For Marsili's library project see above all M.C. Bacchi (2012).

[14]On Marsili's instruments see Bolletti (1751), Emiliani (1979), Baiada et al. (1995).

scientific laboratories, starting with those in Palazzo Poggi. And it was precisely this close link between Istituto and Accademia that attracted the attention of the scholars of Göttingen who had opened a discussion on the renewal of the universities in the German-speaking area that would also influence the proposals that, a few decades later, would serve in the establishment of the university of Berlin.[15] The second was a printshop, that Marsili obtained with considerable difficulty between '18 and '19, and which would provide the Bolognese scientists with an instrument they were totally lacking and which, in the context of the development of the new science, had become indispensable. It meant being able to propagate one's own scientific output as quickly as possible in a world that had made scientific communication an instrument that was irreplaceable in order to avoid being cut off from the rapid flow of information now essential for those carrying out scientific research in the field. And the printshop also played a role above all in the years in which the *Commentarii* were finally born. This was the scientific diary of the Istituto which Marsili had been thinking of since the beginning.[16] The organisation given to this extraordinary and novel encyclopedia of artistic and scientific knowledge, as Corazzi also defined it in his inaugural speech for the Istituto's activities (1714), was destined not only to be long lived but also to consign a picture of Marsili to posterity that was very different from the picture seen by his fellow citizens during his lifetime.[17] The city's gratitude for this enterprise soon began. Already in '39 Scarselli's engraving,[18] which shows the visit by the Elector of Saxony to the Istituto, provides a rather clear picture of Marsili's structure, of the extraordinary results deriving from the proximity and intertwining of the new forms of knowledge. A few years after the description that Bolletti (1751) provided in the volume *Dell'origine e dei progressi dell'Istituto delle Scienze di Bologna*, inspired by the new age of Pope Benedict and by the institution's international launch, ventures even further in praise of its founder who is represented in a blaze of glory that overwhelms all shadows, starting with Breisach.[19]

The Istituto could look benevolently on its recent past: under the protection of the new prince who intended to combine science with faith against the backdrop of the advancing enlightenment, the Istituto's professors grew in authority and had the resources and instruments available suitable for the new design. As we have said the *Commentarii* marked, if slowly, their research journeys, the mesh of censorship loosened, Newtonian philosophy had in the meantime become the most solid reference point for the confirmation and enhancement of the Galileian spirit that had inspired Marsili and the Inquieti. Not only that but female researchers who had already proved themselves in the universities or in other places of research started

[15]Marino (1975); A. Angelini (1993).

[16]On the printshop see J. Stoye (2012).

[17]On the analytical "reading" of the contents of the *Commentarii* see W. Tega (1984), pp. 65–108 and above all Tega (1986, 1987), with contributions by E. Minuz, V. P. Babini, S. Taglianini, N. Urbinati, L. Neri, M. L. Caldelli, P. Gozza, E. Baiada.

[18]A. Angelini (1993).

[19]The bust of general Marsili, due to Piò and placed above the wooden carving of the eagle meant to illustrate the glory of Gustavus Adolphus and belonging to Christina of Sweden, was placed in the entrance hall of the Institute in 1766.

to work in the Istituto's laboratories, something absolutely without precedent in the traditions of the European scientific academies of the age. The most outstanding examples in the Istituto in the age of Pope Benedict were Maria Gaetana Agnesi, a mathematician from Milan, and Laura Bassi, a professor at the city's University; it was the latter, in reality, who marked a real and significant female presence in the Istituto. The role played by Laura Bassi in the advancement of that part of Newtonian physics which concentrated its attention on the electric fluid was far from being of secondary importance: she encouraged a knowledge of the writings that the great physicist from Turin Giambattista Beccaria had devoted to natural and artificial electricism and trained Lazzaro Spallanzani and Luigi Galvani in this new discipline in her private laboratory.[20]

Despite all this it is still remarkable that, on the one hand, this glorification left no room for Marsili's difficult cohabitation with the "mixed government" with which he carried on a long and exhausting confrontation, and on the other hand was destined to have no repercussions in the following century when there would have been every reason for the city to acknowledge its fellow citizen's European stature, scientific ability and cultural foresight. When at the end of the eighteenth century the Istituto delle Scienze was suppressed Marsili's "encyclopedia" underwent its first metamorphosis: a considerable part of the instruments and collections migrated to the laboratories of the University, now finally reformed, while the gypsotheca of the Accademia Clementina, formerly associated with the Istituto, served as a basis for the establishment of the Accademia di Belle Arti (Academy of Fine Arts). But the "heterogenesis" that often governs the life and the function of objects had to make the many finds left in the old Palazzo, including the Antiquities Room, available to relevant sections of the new ordering of knowledge, as called upon by new times. Indeed they were of use to the city when the Country, in the aftermath of unification, began to rebuild its own historical memory. And so the antiquities collection found their place in the Museo Civico Archeologico (Civic Archaeological Museum), the Turkish weapons and trophies enriched the collections of the Museo Civico Medievale (Civic Museum of the Middle Ages), while the Zambeccari collection of paintings formed the first core of the Pinacoteca Nazionale (National Art Gallery). Even so Palazzo Poggi still keeps its memory both in the reassembled scientific character of its rooms and in the considerable nucleus of books and in the collection of manuscripts and finds held in the Biblioteca Universitaria (University Library). The conception of the structure of the world that Marsili's objects sum up was destined to transcend the confines of the century in which it was born. The endless migrations and the learned anarchy of the objects have not succeeded in cancelling the deep trace left in the cultural structures that have marked his city over the past three centuries.

[20]Tega (1969, 1984); on Laura Bassi see Cavazza (2006); but see also Findlen (1993).

1.3 The Homage of the Republic of Letters: Marsili and *Europe Savante*

But the early years of the Istituto's life brought to light both the objective difficulties that such a new and ambitious project was bound to come across in its operational phase, and the oversights and inadequacies of the city authorities, together with the limits of the scientific and technical personnel called upon to lead it. The criticisms, dissatisfaction, the general's outbursts that reached the Roman curia and did not spare the city authorities, gave rise to a series of conflicts that were difficult to settle and destined to last, with brief interruptions, until '26, when Marsili regained the support of the curia and met along the way Prospero Lambertini, destined to rise to the papacy and become the Istituto's munificent prince and enlightened guide.[21] That same year the general decided to proceed with a second considerable donation that would bring new instruments, new collections and above all new books to Bologna, the fruits of the royalties for his *Danubius*.

Marsili thus showed, before leaving his city again, the passion and interest that had inspired him in planning and creating an enterprise destined to have no equal in Bologna for a long time and he did it with the resources acquired this time by a European fame and prestige that his own city was far from wishing to grant him.[22]

In 1721 the disagreements with the Bolognese Senate and with Rome about the staffing and financing that were necessary in order to complete the Institute came to a successful conclusion and Marsili decided that the time had come to undertake a journey, a real one this time, with which he intended to obtain advantageous results for the Institute. The disappointments that had accumulated due to the difficulties that his Institute met with in its scientific arrangements and the differences with the mixed government, convinced him to take a pause for reflection. This however was neither an escape into solitary research nor a retreat of an existential nature. It was rather a diplomatic itinerary to promote not just his project but also his research activity. He left from Livorno for London where he was received with great solemnity by the members of the Royal Society, of which he had been a member since 1691.[23] During his stay, which lasted six weeks, he had the opportunity to have discussions with the most representative men of scientists of that Country whom he had learned to know through their works, such as Hans Sloane, physician and botanist, author of studies on the reproduction and classification of plants and favourite student of the physician Thomas Sydenham who was destined to succeed Newton in the post of President of the Royal Society; with the botanist William Sherard, a convinced Linnaean and pupil of John Ray at Cambridge, who took him to meet the great Newton (then aged 82); with the royal astronomer Edmond Halley who presented him with his *Synopsis Astronomia Cometicae*, with Mead. But he also met two celebrated scholars of the

[21] Angelini (1993)

[22] Angelini (2012).

[23] With regard to Marsili's reception at the Royal Society, according to what J. Stoye wrote, an important role was played by the mathematician and man of science George Ashe, closely linked to Oldenburgh and to his entourage. See J. Stoye (2012), p. 174.

history and theory of the Earth whose doctrines however he did not share, William Derham who had entrusted his devout vision to *Physico Theologia* published in 1713 and John Woodward author of *Specimen Geographiae Physicae* (1704) who followed Burnet's thesis of the effects that the Universal Flood had wreaked on the definition of the structure of the Earth. In any case these meetings had a very positive outcome in that they allowed Marsili to call to mind Malpighi and Guglielmini who had been members of that important Society and to illustrate the progress of the Bolognese Institute which was now ready, and indeed anxious, to continue the collaboration that had begun with such authority. And it allowed the people to whom he talked to urge the General to print, possibly even with an economic contribution from them, the work on the Danube of which advance notice had been given in the Nuremberg *Prodromus* and which was actually dedicated by its author to the Royal Society. But his stay in London could not end for Marsili without a visit to the most famous bookshops at Saint Paul's where he made an appropriate selection of British scientific output to add to the library already donated to the Institute.[24]

On 23 February 1722 he left for Holland where in Leiden he had the opportunity to meet and to become friends with Hermann Boerhaave, then a very famous personage whom the great Musschenbroek did not hesitate to place alongside Newton. Marsili knew and shared not only his approach and methods regarding Natural History and Medicine in particular, but also the Leiden scientist's great interest in a new discipline that was growing on the frontier of several domains of knowledge such as Chemistry. From Leiden Marsili went to Amsterdam to visit the famous naturalist collections (above all of shells) and the city's botanical garden and to obtain material to include in the herbarium that he wanted to donate to his Institute. But his stay in Holland coincided with his official consecration as a natural philosopher of European stature and fame. Indeed it was within the Company of printers working between Amsterdam, Leiden and the Hague that the conditions came to fruition for the edition of his major works that until then had remained unpublished. Negotiations came to a first positive result with the proposal to prepare the edition of the *Histoire Physique de la Mer*, which was indeed published in 1725 with an extraordinary preface by Boerhaave. It was this foreword, together with the commendation that Fontenelle would write in 1730 in the *Memoires de l'Académie des Sciences,* that illustrated the role that *Europe Savante* reserved for the naturalist and the founder of one of the most advanced places for the practice of the new science; but to confirm the saying that no man is a prophet in his own land, this recognition, instead of facilitating Marsili's work in Bologna, made his initiatives more uncertain and arduous, as happened with the events that would accompany the proposed edition of his work on the Danube.[25]

It would have been enough for the Senate and for the Bolognese magistrature to read the foreword that Boerhaave wrote for the edition of the *Histoire Physique de la Mer* in 1725, to arrive at a clear understanding of the European stature that their fellow citizen had also acquired in the field of scientific research. After speaking of the great originality and excellence of the method of a work long awaited by those

---

[24] They are part of the volumes included in the second donation, For this see M. C. Bacchi (2012).
[25] For his relations with Dutch publishers see J. Stoye (2012).

who cultivated natural history, Boerhaave stressed that during his research carried out over fifty years Marsili 'has brought to light a quantity of things that no one had thought of before and that he had arrived gloriously at the discovery of many others that the most astute enthusiast of the subject had barely dared to wish for.' The author was a very special person: 'he had overcome the bitterness of imprisonment by the Turks, imitating the example of Caesar who, when he was taken prisoner by pirates, continued assiduously to cultivate the muses, and found, in misadventure, the really extraordinary consolation of discovering many secrets of nature.'[26] But beyond this he had succeeded in closely interweaving research into natural things with military life, as had Cyrus, Mithridates and Dioscorides before him. Indeed, raised in the school of Mars and having reached the highest ranks in the military hierarchy, amidst the horrors of war he had never ceased to take advantage of every chance to dedicate himself entirely to the contemplation of natural things. He embodied the ideal of Pallas, the goddess who oversaw arts and sciences, who wore the helmet with grace and grasped the spear with decision. Boerhaave declared his own admiration for a man of arms who dedicated his free time not to the typical pastimes of military life but exclusively to the study of nature 'to discover its order, works and laws' to acquire a perfect knowledge of something beautiful, useful and necessary.[27] An example of virtue that had led him to cultivate his sublime genius 'inspiring in him the taste for solid science' in which he had made progress of such usefulness to the present century and to posterity as to make his name immortal and to deserve the eternal gratitude of humanity. 'Whether he was leading his troops or was engaged in other functions connected to his command rank he was always careful to record everything that could be of benefit to the practice of physics, and when he came across anything useful he jotted it down accurately in his notebooks, continuously formulating hypotheses for research and making new predictions to then return to them and examine them and draw benefit from them when the time was ripe.[28] So Marsili, in every favourable circumstance, took notes, collected, selected, memorised what he found most interesting in his observations, which he carried out on various aspects of reality, mistrusting everything and therefore not even completely trusting himself, with the aim of discovering the way in which nature manifested its laws. But there was an absolute novelty that made the observations and conclusions that Marsili arrived at even more precious and that Boerhaave did not hesitate to stress. Until then only the investigations he had carried out 'on land' and that were without doubt worthy of consideration, and even imitated, were known. But this work of his constituted a really important novelty. His research had been concentrated on an aspect still essentially unexplored by philosophers of nature, the sea. On this point, Boerhaave added, this research required considerable resources and so only a farsighted authority could have taken it on, but he added immediately after that, because of the immense labours and dangers that they entailed, they could only have been carried out by a man with the temperament and exceptional qualities such

[26]Boerhaave (1725), Preface, p. I.

[27]Idem, p. II.

[28]Ibid.

as those shown by the author. He had shown himself to be capable of facing the perils of the sea, which were certainly no less than those of war, of bearing, without worrying about safeguarding his own life, all the tranquillity of his spirit in the study of nature, and, in this case, he was not dealing with his soldiers but he had to endure the coarseness and the greed of sailors and fishermen, coax them with promises of money, question them about matters most distant from their daily lives, force them to observe things that they had always despised or to which they had never paid attention. It was at this cost that Marsili acquired the capacity and doctrine to be able to teach others about the essence of things and to make the hidden 'forces' of nature evident. It is thus that, by virtue of his example, he showed what paths he wanted to be followed in search of things of nature and that, with this spirit, he wanted teaching to be done in the Institute that he had founded in his city. 'This Academy, that through his generosity he has built and supplied with such considerable income, has been consecrated by him to the liberal Arts. There has never been anything more glorious. Has there ever been an undertaking more worthy of a King?' But on the merits of our author, Boerhaave concluded in this first part of the foreword, may we be allowed to stop here, 'the fame that will outlive him will entrusted to make them known until the end of the world.'[29]

It must be said however that the Dutch scientist's long introduction did not merely express a series of praises of Marsili's work but also put forward a detailed interpretation of it which, if on the one hand it accentuated his adherence to the post Baconian line of research postulated by Boyle, and therefore explicitly distancing himself from the Cartesian approach of mechanicism, on the other hand insisted on the affinity between Marsili's text and his own research in chemistry and medicine.

So first of all there is the method that proposed a close relationship between senses that do not deceive us and reason that does not hem us in within the confines of deductive knowledge.

'Man has received from the very hand of the creator of all things the faculty of knowing some particular properties of things which, however, remain under the dominion of the senses; and this is a truth so established that neither Pyrrho nor Socrates would dare to overturn it in doubt. Beyond this all wise men concur regarding the great usefulness that the knowledge of objects and practical activities have for everyday individual and social life, as is demonstrated for example by astronomy, medicine, navigation. There are two paths through which all our knowledge passes. The first is guided by the activity of the senses, the second proceeds from reasoning and reflection. But in the end it is inevitable to conclude that knowledge does not make any progress if the one does not assist the other, so it is therefore necessary to conclude that it is only when Art, that is to say the initiative of man, operates so that they converge that they will produce those extraordinary effects that they are capable of. Only if these two paths are firmly interwoven will true scientific knowledge of the world be arrived at.'[30] And Boerhaave stressed with evident satisfaction Marsili's Baconian and Boylian inclinations when he stated that 'it is necessary to respect

[29]Idem, p. III.

[30]Idem, p. IV.

rigorously this inviolable order according to which the knowledge of bodies that one acquires through the senses must proceed first, since it must be brought into use and practiced well when reason launches its discussions' and it has been amply confirmed, indeed, that the more reason is preceded by the experience of the senses, the more the spirit is shown to be ready and fertile at the level of the knowledge of bodies and their relationships. The task of reason therefore will be eminently that of 'debating, examining, conciliating the various phenomena that experience has gathered, so that it can only be concluded, regarding bodies, that what it has known with the most extreme clarity, is none other than a necessary consequence of the facts that the senses have caught sight of there.'[31] And this state of affairs has an immediate consequence - that the bodies which one intends to investigate must be within reach of the senses: 'it is for this reason that, when one wishes to have the pleasure of observing far away objects, one must begin by drawing near to them. Even today – Boerhaave added polemically turning his thoughts to his *Sermo Academicus de Comparando Certo in Physicis* which he had read in Leiden in 1715—there are few that follow this path and the reasons that impede or delay the progress that physics could make lie precisely here.'[32]

Indeed Boerhaave appreciated and actually emphasised this linear process which from the senses (appropriately directed) led to reason (of appropriate scope) since it was this that constituted the true added value, in other words the excellence of the method adopted by in the *Histoire*; it consisted in the fact that his observations took into account the individuality of the bodies and the phenomena and, above all, they took place in the very place where the bodies and phenomena originated. That is to say they could be examined before they had the time to degenerate or substantially change their state compared to their initial condition, to their *statu nascenti*. 'These are observations that, unlike all others—Boerhaave insisted–really enrich physics providing the natural philosopher with the opportunity to resort to them safely, whenever he is seized by the desire to meditate on or write about natural things'.[33] It is still these observations that make us see the inestimable value of the *Histoire*; they teach us to give due weight to their excellence compared to all others.

But it must be said that the criticism of the old mechanicism was not limited to only this argument. The experimental data that Marsili arrived at in his work confirmed that 'the whole globe of the Earth that we inhabit is really an organised body.'[34]

As we have said it is precisely on his reflections on this concept of an organised body that Boerhaave concentrated all his experience as an anatomist, physician and chemist; in this way he sought further accord with Marsili's intentions when he faced the need to confront, all the way, the essence of the single bodies that constituted the 'regulated structure of the Earth', down to the reading of the infinitely small, that is to say that which escaped even the most careful observation of bodies but which could not remain hidden either to the eye multiplied by the microscope or by the processes

[31] Idem, p. V.

[32] Ibid.

[33] Ibid.

[34] Idem, p. VI.

of analysis and decomposition adopted by chemistry. So the machine the Earth is an organised body not only in the sense that it is made up of an enormous number of different parts. Indeed each of these parts has two simultaneous tasks: the task of carrying out its own functions and the task of contributing with their efforts to the achievement of the greater work: the total and ordered movement of the globe which can only be attributed to the dynamic harmony that governs its parts.[35] Certainly one cannot infer from Marsili's text 'that these parts are united by chance and that they give rise to an improvised whole, confused and without order. On the contrary it is a certain proof that it was only the creator spirit of all things, whose wisdom is infinite and whose power is without limits, who formed them all and arranged them in such a way amongst themselves, that all their operations tend towards a single common aim. All things created on this Earth are of such a nature as to be able to last until the end of centuries and to have during this time their vicissitudes, their revolutions, their regenerations in accordance with a certain law, a determined order.

But Boerhaave's reflections seem to move beyond the medical and chemical horizons to arrive at a series of considerations that are set on that level of that history of the Earth. Here the reasonings of a time different from biblical time and the always provisional conclusions of a movement that brought nature to life, and that therefore could not be attributed to the only form of movement that physicists and geometers took into consideration which consisted of moving a body from one place to another, were woven together. In this world, where chance is rigorously excluded,[36] within this cyclical nature, whose rhythms and characteristics Boerhaave stressed so as to make it resemble ever less a machine and ever more a living being, this again invoked the physician rather than the physicist, 'all things have their birth, their duration, their changes; they die and are reborn with the contribution and with the help of all the others, that gather together and constitute the Earth'.[37] An image of nature emerges that has lost every relationship not only with the image of the flood, but also with the mechanicistic or corpuscularian one that reduces matter to figure and movement. A nature made up of parts equipped with their own dynamic force and an attractive and repulsive capacity; that is a survivor of its history of catastrophes to which it owes its form which is far from being defined once and for all; which tends to put itself forward to man's knowledge as a continuous chain of bodies and phenomena; that contemplates a series of relationships between its parts. A nature in which the events of the origin and dissolution of bodies acquire a cyclical dimension that displays a dynamic order that characterises it, starting from its very origins. So no credit is given to the theory of original chaos to which Cartesian doctrine referred nor the evocative image of the clockmaker god.[38] The laws of nature, as Boyle and Newton claimed, are part of the very occurrences of the world, they emerged from the hands

[35]Ibid.

[36]Ibid.

[37]Ibid.

[38]Boerhaave insisted on the question of original chaos that Marsili denied at its roots. On this question, which by now had become the strong suit of all the criticisms of Cartesian doctrine see the observations of Boyle, Leibniz and Newton. For this see above all Rossi (1986).

of the creator during the act of forming the Earth and the planets, we are only allowed to know them through observations and measurements.

We are convinced that one must no longer attribute to chance that which happens everywhere on Earth. Besides, to obtain a confirmation of this order, of this agreement, of this reciprocity and understanding as the cooperation of its parts in the realisation of a common and general design, it is sufficient to observe what happens in the human body. So around the anatomist's table Boerhaave also summoned the chemist and the conclusions these 'interpreters' of nature arrived at showed several analogies with the conclusions that Diderot, attracted by the living forms of nature that change with sudden inexorability, had put forward in his *Pensées sur l'Intérprétation de la Nature* that began significantly with a very Baconian epigraph: 'man is not a machine, nature is not God, a hypothesis is not a fact'.

'The eye of man, to be able to exist as a whole and to be able to carry out its functions, cannot do without, before all else, the joint operations of mouth, throat, ventricle, intestines, veins, heart, lungs, arteries, brain, nerves and all the other organs. I am convinced that when the works of nature are considered from this aspect and when they are arranged in good order we will have a more exact idea of nature than that which is commonly accepted.' [39]

As for the exposition of the contents of the *Histoire*, Boerhaave did not depart from the assumptions of Marsili's text and he stressed above all what distinguished Marsili's work from the alleged explanations of the history and structure of the Earth by the British and continental diluvians, starting with Scheuchzer. And regarding this distinction he again seemed to wish to emphasise it when he highlighted the significance that Marsili had chosen to attribute to the absolute structural and physical compatibility between the bottom of the sea and the mountains that emerged from the waters, and also to the significance in the reconstruction of a real history of the Earth, of the layers that visibly constituted it and the fossils that inhabited it, and finally to the specificity of the plants and animals with special reference to the presumed ancipital nature of coral.[40]

But, briefly having summarised the arguments dealt with in Marsili's study, Boerhaave again stressed the originality of the method that the author of this great work had chosen to adopt. He had not followed the opinions of others and he had moved beyond the circle of prejudices, going to observe and examine things with his own eyes, he had sought and observed in those very places everything that he described. He had cited his sources rigorously, and he had separated those that were fully reliable from those of uncertain value and from 'romances'. He had carried out research in the field, accompanied by fishermen and, not least obviously, he could claim on this matter forty-five years of observation.

[39] Rossi (1986). On the 'crisis' of the mechanicistic concept of nature see Tega (1971).

[40] Marsili (1726). As well as his expertise as a physician and chemist, in the second part of this *Preface* Boerhaave shows his great interest in a history of the Earth from the antidiluvian and, at heart, antimechanicistic standpoint. For information about him, as well as the already cited work by F. Abbri; see Lindeboom (1968); but now in particular Powers (2012).

'In other words a philosopher who does not stay in his laboratory, but on the sea, far from people of letters, alone amongst seafarers, far too from the silence and tranquillity but amidst tumult and clamour, not amongst the comfort of peace but amongst the alarms and frequent perils of life. Take benefit from this book that makes available to you the fruits of an extraordinary work that you cannot hope to find elsewhere'.[41]

Evidently Marsili's research met with the favour of cultured opinion in the United Provinces. The editors too, encouraged by Boerhaave himself, urged the General to send from Bologna the manuscripts of the great work on the Danube whose *Prodromus*, published in Nuremberg in 1700 and dedicated to the Royal Society, had received the praises of the most famous natural philosophers. Faced with the proposals from the Dutch editors, Marsili abandoned the idea of having his work printed in Bologna and in March 1723 he reached an agreement, or rather he signed a contract with the company of editors of the Hague. The agreement contained several clauses regarding the return of the manuscripts to the Institute once printed, and a ban on publishing the work again for the next hundred years; Marsili should have received a payment of ten thousand Dutch florins but the author renounced this on condition that the Company replaced the sum with books of the same value, destined for the library of the Institute. The list of preferences would be sent to Bologna for revision and returned again to the editors.

It was the scientists of the Royal Society and even more so the scientific milieux of the United Provinces led by Boerhaave, who decreed Marsili's consecration as man of science, or rather natural philosopher, and it was the editors who decreed his launch into the *République des Lettres* when they placed by his side a Huguenot journalist from Utrecht, De Limières, who was commissioned to write a booklet to describe the figure of the natural philosopher and founder of the Institute of Science.[42]

After his return to Bologna and having solved the question of the legitimacy of the use of his manuscripts raised by the magistrature appointed to the Institute, Marsili worked intensely on arranging the impressive material to send for publication which included both the *Histoire* and the *Danubius*. 'The Count –Boerhaave wrote in 1724– is drowning in his Danube'.[43] Indeed Marsili was busy not only arranging the material that he had been able to overhaul with the wise Muller and the Dutch engravers, but striving to enrich some points that he considered crucial for the work that were linked to the research and observations that he had been able to carry out while dust was accumulating on the old manuscripts. As he concentrated on expanding the parts regarding the Roman presence in the Balkan region and in Dacia, on the Tiberius bridge, and on the Roman communications network, he entrusted to Antonio Rossi, a pupil of the Clementine academician Franceschini, the completion of the engravings regarding the cartographic part of the work and he asked Morgagni to check and correct his notes on the anatomy of the sturgeon, this time expanded on the basis of the specimens of the same species coming from the Po. At the end of 1724 the six

[41] H. Boerhaave (1725) *Preface*.

[42] De Limières (1723).

[43] J. Stoye (2012).

volumes were in the hands of the editors who concerned themselves with the title pages and the engravings that should accompany them.[44]

Having brought to an end this hard work of editorial fine tuning, Marsili undertook a journey of pleasure and rest on Lake Garda, but the demon of research was inexorable; even here he made observations that would then be summed up in the work *Osservazioni fisiche intorno al lago di Garda detto anticamente Benaco* (Physical observations on Lake Garda in the past called Benaco).[45] In reality Marsili had withdrawn to Maderno, on Lake Garda, to conclude his essay *Sullo Stato Militare dell'Imperio ottomano* (On the military state of the Ottoman Empire) which should have concluded his critical narration of the knowledge acquired during military campaigns, of his diplomatic relations and visits to Constantinople, but he had to admit that the 'conversation amongst Turks was made so difficult and harsh for me that for my relief and amusement I felt necessary some tamer and more charming study from time to time'. Distancing himself from the memory of military events that had brought him into close contact with the Ottoman world would first drive him to write a dissertation that he had intended to send to his friend Boerhaave about some physical observations that he had made during the recent journey that had taken him from Livorno, through the straits of Gibraltar, to the English Channel which was overlooked by France, England and Holland and then to a study of the lake that he saw from his windows every day and was the place of his walks. In other words the Baconian curiosity overtook him again and with it the irresistible attraction of naturalist observations. And he did this with vigour and with good disposition 'not so much because the subject was so obvious to me, as it was, *but because I saw it as proper and befitting to my intent to prove the organic structure of the Earth, on which for so many years I have undertaken a collection of observations examining mountains, plains and seas.*[46] Thus I imagined this lake as a small sea that for so many probable reasons should also have a solid internal structure corresponding to that of the seas, since this like the other lakes are not accidental but necessary for the good regulation of the watery mass that flows over the Earth and will in any case flow until the destruction of the Earth in the same way as the seas. But when finally in 1727 the *Danubius* and the work of De Limières reached their readers they would become intertwined with a further series of accolades that Marsili obtained in Paris, thanks to the intervention of Abbé Bignon, mentor and founder of the Académie des Sciences of which Marsili was a member, and thanks to the intervention by Fontenelle who was its President, an outstanding distinction, that is to say the praise that was published about him posthumously in the *Mémoires* of that Academy.

For that matter, as had happened for Malpighi, Montanari, Guglielmini and Morgagni, the city did not understand either the magnitude of the credit that Marsili was receiving in Europe nor the international status that had been acquired by the Institute that he had so greatly wanted to see born. After the dispute, resolved with

[44]On the relationship Marsili maintained with the Clementine Academy and on the Clementine Academy in particular, see Benassi (1988).

[45]The text edited by Marsili in 1727 was published by Longhena (1930).

[46]For the articles on Benaco and on Garda see. M. Longhena (1930).

extreme difficulty but which must have constituted another considerable advantage for the Institute, others followed and only the protection of the Roman curia and the appearance on the scene of Prospero Lambertini succeeded in quelling new conflicts already on the horizon and in propitiating Marsili's second donation which thus allowed the Sir d'Aquino (here reference is made to Tommaso d'Aquino), who had abandoned many of the expectations placed on the Institute, to conclude his last journey in accordance with the motto 'Nihil Mihi', which when you think about it had been the distinguishing feature of his entire existence.

1.4 Marsili's legacy and the princedom of Pope Lambertini

The period that Prospero Lambertini spent in Bologna as archbishop (1730–1741) was characterised by a formal and discreet relationship with the academicians which aimed to favour in every way the growth and consolidation of the new scientific institution without offending the susceptibility of the Bolognese Senate, of the University and of the Holy See. However he played his most important role after he ascended to the papal throne with the name of Benedict XIV. From that new point of observation it immediately appeared clear to Pope Benedict how the Institute of Science and its Academy, better than other institutions that had repeatedly failed in this aim, could restore to Bologna that eminent position in European culture that it had by then long since lost, how an alternative of this kind, right in the cultural heart of the Catholic world and at the urging of the pontiff himself, would have allowed a space for research to be kept open within which the conditions could be created to recreate the relationship between faith and science which had been severely damaged after the condemnation of heliocentrism.

This aspiration, openly declared many times, led Lambertini to a gradual re-evaluation by the Church of those scientific-philosophical concepts that had belonged to Bacon, Galileo, Boyle and Newton and that had considerably influenced the activity of the Bolognese scientists between the Seventeenth and Eighteenth Centuries and placed their concrete initiative on a soil that actually would lead him to achieve, enlarge and progressively innovate, the enlightened design of European scope conceived by Ferdinando Marsili.

The relationship and alliance between faith and science were difficult, especially given the climate of mistrust and sometimes hostile closure that the ecclesiastical hierarchies still showed on this score in the mid 18th century; but they were a necessary relationship and alliance if one wished to relaunch the role and eminence of the Church in a European horizon in which, by then, the results of the new science were being asserted and the examination of holy texts and theological doctrines was becoming ever more free and flagrant. But it was equally clear to Benedict XIV that the Institute of Science could answer this aim, produce significant examples in the direction to follow, if in the meantime it managed to establish itself as the authoritative chair of the new science, as a place of research whose results were awaited and read with attention by the international scientific community. This provided a series of further valid reasons to encourage a pope already well disposed to put into play all the necessary intercessions to turn the recent Bolognese institution into a strong point of scientific culture which by then was prospering under the clear and exclusive influence of Newtonian science.[47]

And his attention towards the Institution and the Academy was assiduous and logical, an attention not lacking in impediments and disappointments but in the end also crowned by successes that Benedict recorded and stressed with open satisfaction. The Pope was not unaware that the crisis of the university had its roots not so much, and not only, in the cultural and social situation current in the city and throughout the Papal States but rather in the profound changes that had occurred with regard to organisation and research in the European context. Italian universities and institutions had not known how, or been able, to take these into account, partly because of the interference of ecclesiastical censorship. Lambertini's judgement with regard to the customs and habits of the University's professors was always severe and pungent but it became harsher and more concerned when he perceived the concrete risk that the worst vices of the ancient University might migrate to the Institute and hamper its activity. But his reproofs and urgings did not bring with them any underestimate of those who contributed with virtue and constancy to the growth of the new undertaking: "great is our consolation—he wrote to Cardinal Magnani in February 1746—in hearing that in the Institute the Academies proceed. This is as much as can be hoped for in Bologna regarding literature (science) since the case of *ius* and good theology is desperate".

It seemed pointless, and even harmful, according to the Pope to cultivate mistaken illusions: it would not be thanks to an improbable rebirth of the "legal subjects", but in virtue of the physical sciences and of the activity of the Institute that "our common homeland will again take on the true title of mother of universities in the conception of men of taste". In Benedict's reforming intervention the agreement with Marsili's paradigm was obvious: it would be the experimental sciences together (Medicine, Chemistry, Natural History, Physics of fluids, Astronomy) and Mathematics that would restore to Bologna a place of eminence in European culture that had once belonged to Law, Philosophy, Theology.[48]

The importance assumed by the Institute in the 1740s no longer left room for extemporary interventions and support by benefactors. A new library, anatomical waxworks for doctors and artists, equipment for the astronomical observatory, material for the study of Natural History and Chemistry, machines and laboratories for the study of electrical fluids and for the experiments on light, publications to allow the Institute and the Academy to communicate with the network of scientific institutions that in the meantime had been established in the European context – all were necessary. It was a question of redesigning Marsili's Institute and this required the direct intervention of the Pope whose assiduous attention was greatly appreciated by the academicians. Benedict made himself the protagonist of a reform that concerned the very management of the Institute with the aim of bringing to an end the climate

[47]Tega (2011).

[48]Ibid.

of mistrust and squabbling that had become established between those who were the protagonists of the scientific activity and life of the Institute, such as the professors who should be guiding the laboratories, and those who had inappropriately seized control and despotic direction such as the Institute's Assunteria (administration). Until Benedict's reform (*Motu proprio* or decree dated 22 June 1745) the financing of the Institute was entrusted to the Senate and to sporadic episodes of private generosity. Benedict, who knew Bologna's political class well, recognised the urgent need to guarantee the Institute not just economic stability but also financial autonomy by allocating to it the revenue of the former Collegio Panolini that not even the city's Senate could alienate. To prevent the Institute following the rather unfortunate fate of the University the service of showing experimental science to the public had to be improved and its research capacity expanded. So it was necessary to have a limited number of professors who were curious, competent, dedicated to research and remunerated in a way that did not differ that much from physicians and jurists who were favoured from this point of view by the profits from private practice.

Papa Lambertini knew well that the European fame of Marsili's Institute was entrusted to the activity of the Academy and its scientific diary, the *Commentarii*, which broadcast the results of research carried out in the Institute's laboratories to the European Republic of Letters. For this reason the structure and working of the Academy also had to be reformed: the collaborations had to be regulated, the members' tasks had to be redefined and the annual pensions to the 24 academicians (called Benedictines) most directly involved in experimental scientific activity had to be fixed. But it was also necessary to stem the tendency towards provincialism and isolation by encouraging contacts, correspondence, scientific agreements and programmes capable of forging links between the Bolognese institution and other outstanding institutions such as the Académie in Paris, the Royal Society in London and others in Leipzig, Berlin, Saint Petersburg.

And it was also due to Benedict's direct intervention and the influence that he long exercised that Laura Bassi was appointed a supernumerary amongst the Benedictines and Maria Gaetana Agnesi was offered the chair in Mathematics and that Manzolini was given the task of continuing her husband's work and realising the most modern part of the anatomical waxworks, acquired by the Institute and praised without reserve by Galvani after her death. Voltaire, M.me du Châtelet, Clairaut, Maupertuis, Buffon, d'Alembert, Formey who certainly were not close to the doctrines favoured by the Holy See were called upon as "foreign academicians" and amongst those associated with Italian Academies or Universities there were Boscovich, Paolo Frisi, Giambattista Beccaria who made no secret of their belief in the physics of Galileo and Newton.

Pope Benedict followed a double strategy: that of establishing a place destined to establish a relationship between faith and science at a time in which the language of European culture was giving substance to a new phase of science of man, of nature and of society and, as he confessed in 1746 to his friend Paolo Magnani, to give an endowment to the Institute, to put it in a condition to make Bologna famous as the University once had done.

Two tasks entrusted to a group of scientists who, just before the pope's death, published the first volume of the *Commentarii* which reinforced throughout the Republic of Letters the experimental direction clearly inspired by the figure of Newton for natural and experimental philosophy.

1.5 The '*Scientific Diary*' of the Institute and the Academy

At this point it is worthwhile to refer to the work that accompanied the experimental activity of the Academy members until the end of the century, that is to say the *De Bononiensi Scientiarum et Artium Instituto atque Academia Commentarii*—(the others were published by the editor Lelio della Volpe, in the years 1745–47, 1754, 1757, 1766, 1780, 1791, respectively) which together provide a faithful enough diary of the scientific activity of the Bolognese academicians up to the dawn of the new century. The material was arranged in three parts: the first reported in detail the origins of the Institute and of the two Academies that operated within it, of its institutional structure, of the establishment and the growth of its equipment and instrumentation.

The second part, gathered together under the general title *Commentarii,* occupies a central place in the organisation of the whole volume; here the secretary reported not only the essential themes tackled in the academicians' dissertations but justified the discussions opened up around them as well as the theses that were discussed on each single question in the scientific community.

It was a question of painting a picture of the activity of an institution which was young but now fairly complex and this was one of the tasks, I think the most important one, that the secretary was called upon to carry out. He became a key figure, bringing together a series of extremely delicate duties, the decisive coordinator for some crucial aspects of the Academy's life: the general definition of the calendar of activities, the organisation of the single work sessions and laying out the research results and the discussions to be printed.

Francesco Maria Zanotti, who carried out the role of secretary in exemplary fashion in one of the most intense periods of the Academy's life, already in the first volume of the *Commentarii* offered an exemplary example of scientific diary and arrangement of material which would not later undergo substantial changes.

The secretary's fluid and knowledgeable pen put forward from the very first pages the persons, questions and debates that animated the city's scientific milieux; he let academic rivalries and contrasts between different generations of scholars and researchers be understood clearly; he told of lively disputes in which he took on the task of neutral arbiter; he diligently highlighted and enhanced all those aspects that linked the debates and research of this small scientific community to the wider Republic of Letters operating in the international context.

And the results of these scientific labours, of these quarrels, of this complicated network of relations, are laid out in the second volume; an order that Zanotti himself hastened to present as the least incomplete amongst those proposed by contemporary science which, moreover, acknowledged not having succeeded in proposing a truly all encompassing arrangement of knowledge without limits and errors. The arrangement of the material, in the intentions of the secretary and the academicians, should avoid, in any event, both the path of invention and the path of a rigid and binding systematic approach.

And so the Bolognese academicians' encyclopaedia of science started with Natural History to then pass successively in order to Chemistry, Anatomy and Medicine, Physics, Geometry, Mechanics, Arithmetic, Astronomy, reflecting the strategy first thought of for the Institute by Marsili and then by Benedict XIV.

It was an arrangement of the sciences that was not rigid but ordered enough, which proceeded from the particular to the general, from the concrete to the abstract, from the inanimate to the animate, respecting the ascending and harmonious order according to which, to many scientists of the time, bodies and natural functions appeared to be arranged and that benefitted from a regulating but still solid concept of the great chain of being.

The third part of the *De Bononiensi* gathered together the most noteworthy contributions presented by the academicians during the public sessions, chosen and prepared for printing by the secretary with the help of a select commission of experts.

The *Commentarii* certainly did not exhaust the wide picture of the city's scientific production but they were a faithful reflection of it.

For natural history they included contributions by Giuseppe and Gaetano Monti; for the physical-chemical sciences those of Beccari, Galvani, Manfredi, Zanotti, these latter also protagonists of the Institute's intense astronomical activity; for the medical sciences those of Morgagni, Valsalva, Molinelli; for mechanics, hydraulics, mathematics those of Guglielmini, Laura Bassi, Riccati, and Canterzani who succeeded Francesco Maria Zanotti in the role of secretary of the Academy.[49]

Now a careful reading of the *Commentarii* allows us to detect how, when it comes to the influences relative to the lines of research and discussion implemented by the Academy, Isaac Newton's princedom corresponded almost symmetrically to the princedom of the Pope.

Indeed one can say that it was Benedict XIV's urgings that gave a decisive push to the launching of the experimental sciences and to increase the importance of the Newtonian synthesis in the activity of research and discussion that characterised the newest of Bologna's scientific institutions.

For the Institute to flourish and for it to be fully placed in the framework of international scientific relations both the merging that occurred between the Bolognese scientists and the most significant interpreters and supporters of Newtonian physics then available in Europe, and the close collaborative relationship with analogous and often more important scientific institutions in other countries, and above all with the Royal Society in London and with the Académie des Sciences in Paris, were decisive.

It must be said that once again it was the Pope who urged the Bolognese academicians to nominate Voltaire, Maupertuis, d'Alembert amongst the corresponding members of the Academy and to acquire the most assiduous collaboration of the

[49]On the organisation, structure and distribution of the *Commentarii,* see Tega (1986, 1987); A. Angelini (1993).

best known and most advanced representatives of Newtonian physics research in Italy, such as Giambattista Beccaria from Turin, Paolo Frisi from Milan, Ruggero Giuseppe Boscovich from Ragusa, an unchallenged authority of the Roman College. Moreover it was actually in the Institute's golden age, from 1730 to the end of the 1770s, that the network of scientific relations and alliances with European institutions became closest, especially with those mentioned above.

Newtonian science, which left a deep mark in the history of the Institute in the second half of the century, already had many supporters amongst the academicians in the 1730s.

From Manfredi to Galvani the methodological and scientific orientation contained in the *Principia* and in *Opticks* constituted the common reference point for research and debates in Astronomy, Mechanics, Optics, Mathematics, Anatomy, Electrical Physics on which the lines of research of the Institute concentrated with results that were sometimes also of absolute distinction. If, for the first generation of academicians (Eustachio Manfredi, Francesco Maria Zanotti, Jacopo Bartolomeo Beccari), the Newtonian synthesis represented an arduous conquest, a liberation from the old Cartesian models, and from the censorship that was still in operation, that was not easy, those who followed them from the middle of the century (Laura Bassi, Leopoldo Caldani, Sebastiano Canterzani, Luigi Galvani) compared themselves in new ways and measured themselves with Newton's legacy with remarkable results and with the development of new fields of research which the development of Newtonian theory had given rise to.[50]

In any case towards the end of the century the Institute had available a series of laboratories, workshops, and museums of extraordinary breadth and of great interest. It is difficult to say—as was noted by numerous travellers and scholars for whom Bologna and the Institute constituted an obligatory stop on their Italian tour – whether they really found themselves facing the house of Solomon realised; certainly in Bologna not all disciplines were cultivated to European levels, equally certainly the instrumentation of the laboratories and the endowment of the rooms were not always state of the art, and yet those who concerned themselves *ex professo* with Physics, Chemistry, Astronomy, Medicine, Natural History, could find reflected and represented in the Rooms and Library of Palazzo Poggi the map of contemporary science with precise indications of all its territories. The activity of the Academy and of the Science Institute restored to Bologna a place of consequence in European culture. But this time the network of international relations was not sustained by the disciplines of Law, Literature or Theology. This task was carried out fairly brilliantly by Natural Philosophy and the Physical-mathematical Sciences, which found the most propitious conditions for their growth in the Science Institute.

Figures 1.1, 1.2, 1.3, 1.4, 1.5 and 1.6 show portraits of Luigi Ferdinando Marsili and other related images.

---

[50]On the influence in Bologna of Newtonian thought see Casini (1983); W. Tega (1984, 1987), pp. 9-35.

Fig. 1.1 Antonio Zanchi and Antonio Calza, *Equestrian Portrait of Luigi Ferdinando Marsili*, Biblioteca Universitaria di Bologna, inv. 117

Fig. 1.2 Anonymous, Luigi Ferdinando Marsili (1658–1730), founder of the Istituto delle Scienze. Oil on canvas, first half of the 18th century, Biblioteca Universitaria di Bologna, inv. 76

Fig. 1.3 An 18th-century engraving of Palazzo Poggi, seat of the Istituto delle Scienze

Fig. 1.4 The laboratories of the Istituto delle Scienze at the Palazzo Poggi during the visit of Prince Frederick Christian of Poland, 1739. Detail from a miniature by Antonio A. Scarselli, Anziani Consoli, Insignia vol. XIII, c. 140, Archivio di Stato di Bologna

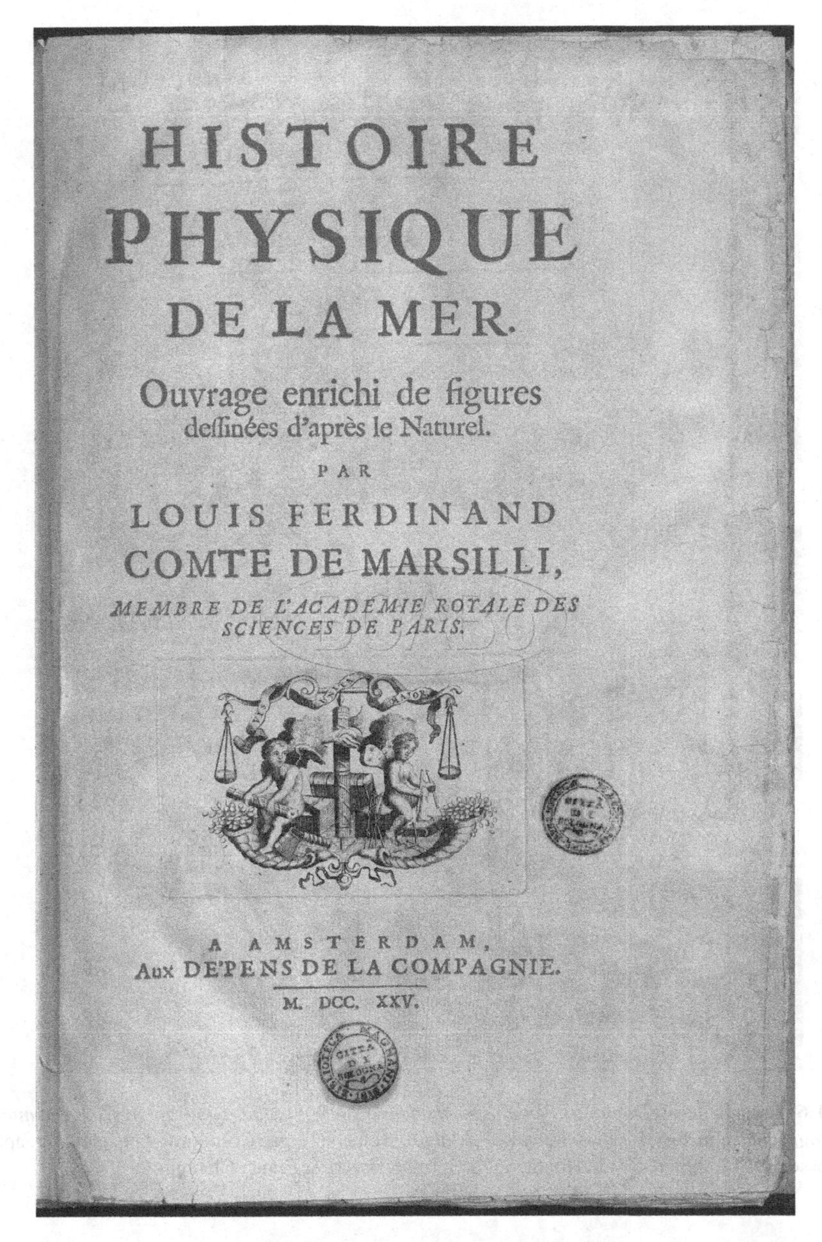
HISTOIRE
PHYSIQUE
DE LA MER.

Ouvrage enrichi de figures
dessinées d'après le Naturel.

PAR

LOUIS FERDINAND
COMTE DE MARSILLI,

MEMBRE DE L'ACADEMIE ROYALE DES SCIENCES DE PARIS.

A AMSTERDAM,
Aux DE'PENS DE LA COMPAGNIE.

M. DCC. XXV.

Fig. 1.5 Frontespice of *Histoire Physique de la Mer*, which Marsili published in Amsterdam in 1725, Bologna, Biblioteca Comunale dell'Archiginnasio

Fig. 1.6 Frontespice of *Danubius Pannonico-Mysicus observationibus geographicis, astronomicis, Hydrographicis, historicis physicis,* by L. F. Marsili , Tome I (Hagae Comitum- Amstelodami, apud P. Gosse, R. Chr. Alberts, P. de Hondt, apud Herm, Uytwerf & Franç. Changuion, 1726)

References

A. Angelini, *L'Istituto delle Scienze e l'Accademia. Anatomie Accademiche III,* Il Mulino, Bologna 1993

A. Angelini, "*Con l'elmo in testa e la lancia in mano*". *L'architettura del sapere di Marsili,* in *La scienza delle Armi. Luigi Ferdinando Marsili 1658–1730,* Pendragon, Bologna 2012, pp. 19–35

M.C. Bacchi, *Contributo allo studio della libreria di Luigi Ferdinando Marsili,* in *La Scienza della Armi. Luigi Ferdinando Marsili 1658–1730,* Pendragon, Bologna 2012, pp. 201–231

E. Baiada, F. Bonoli, A. Braccesi, *Museo della Specola,* Editrice Compositori, Bologna 1995

S. Benassi, *L'Accademia Clementina. La funzione pubblica. L'ideologia estetica,* Bologna 1988 (n.e., Bologna 2004)

G.G. Bianconi (Editor), L.F. Marsili, *Alcune lettere inedite del Generale Conte Luigi Ferdinando Marsili al Canonico Trionfetti per la fondazione dell'Istituto delle Scienze,* Sassi nelle Spaderie, Bologna 1894

Boerhaave, H. in L. F. Marsili, *Histoire Physique de la Mer,* aux de'pens de la Compagnie, Amsterdam, 1725

G.G. Bolletti, *Delle origini e dei progressi dell'Istituto delle Scienze di Bologna,* Lelio della Volpe, Bologna 1751

E. Bortolotti (Editor) *La Fondazione dell'Istituto e la Riforma dello Studio, Memorie intorno a L.F. Marsili,* Zanichelli, Bologna 1930

P. Casini, *Newton e la coscienza europea,* Bologna 1983

M. Cavazza, *Una donna nella repubblica degli scienziati: Laura Bassi e i suoi colleghi,* in *Scienza a due voci,* edited by R. Simili, Olschki, Firenze 2006, pp. 61–85

H.P. De Limières, *Histoire de l'Académie appellée de l'Institutt des Sciences et des Arts, établi à Boulogne en 1712,* Amsterdam, s.e., 1723

A. Emiliani, *Un modello museografico per i materiali dell'Istituto delle Scienze,* in *I materiali dell'Istituto delle Scienze,* Clueb, Bologna 1979, pp.121–138

P. Findlen, *Science as a Career in Enlightenment Italy: The Strategies of Laura Bassi,* in «Isis», 84, 1993, pp. 441–469.

R. Gherardi, *Il 'Politico' e 'altre scienze più rare' in due inediti marsiliani del primo Settecento,* in «Annali dell'Istituto storico italo-germanico in Trento» I, 1975, pp. 85–141.

J. Hartley, *Catalogus Universalis librorum in omni facultate, linguaque insignium*...Hartley, Londini 1699

G.W. Leibniz, *Bibliotheca Unviersalis Selecta,* Berlin, 1689.

G.A Lindeboom, *Herman Boerhaave. The man and his work,* London, Methuen & Co, 1968

M. Longhena in *Scritti inediti di Luigi Ferdinando Marsili,* Bologna, 1930

L. Marino, *I maestri della Germania. Gottingen 1770–1820,* Einaudi, Torino 1975

L.F. Marsili, *Danubius, Pannonico-Mysycus, observationibus Geographicis, Astronomicis, Hydrographicis, Historicis, Physics perlustratus et in sex tomos digestus,* Haegae Comitum apud, P. Gosse, R Chr. Alberts, P.De Hondt; Amsterdam, Apud Herm Uytwerf & Franç Changuuion, 1726

G. Naudé, *Advis pour dresser une bibliothéque* (1627), Paris, 1627

J.C. Powers, *Inventing Chemistry. Herman Boerhaave and the Reform of Chemical Art,* The University Chicago Press, Chicago and London, 2012

P. Rossi, *La storia della scienza e l'emergenza dei problemi,* in *I ragni e le formiche,*Bologna, 1986

J. Stoye, *Marsigli's Europe 1680–1730. The Life and Times of Luigi Ferdinando Marsigli, Soldier and Virtuoso,* Haven-London 1994. It. Edition, *Vita e tempi di Luigi Ferdinando Marsili,* edited by F. Simoni, Pendragon, Bologna 2012

M.Talman, *Elenchus librorum orientalium manuscriptorum, videlicet Graecorum, Arabicorum, Persicorum deinde Hebraicorum ac antiquorum Latinorum...,* Susannam Christinam, Matthaei Cosmerovij viduam, Viennae Austriae 1702

W. Tega *Mens agitat molem. L'Accademia delle Scienze di Bologna,* in *Scienza e Letteratura nella cultura italiana del Settecento,* a cura di R. Cremante e W. Tega, Bologna 1984

W. Tega, *Le "Institutiones in physicam experimentalem di Giambattista Beccaria"*, in «Rivista Critica di Storia della Filosofia», 1969, pp. 179–211

W. Tega, *Meccanicismo e Scienze della Vita nel Tardo Settecento*, in «Rivista di Filosofia», LXII, 2, 1971, pp. 155–176

W. Tega, *Papa Lambertini. Una lucida visione dei rapporti fede-scienza* in A. Zanotti (Editor) *Prospero Lambertini. Pastore delle sua città, pontefice della cristianità*, Minerva, Bologna 2011, pp. 161–170

W. Tega, *Anatomie Accademiche I. L'Enciclopedia scientifica dell'Accademia delle Scienze di Bologna*, edited by W. Tega, Bologna, 1986

W. Tega, *Anatomie Accademiche II. L'Enciclopedia scientifica dell'Accademia delle Scienze di Bologna*, edited by W. Tega, Bologna, 1987

Chapter 2
Newton's Legacy: An Open Field of Research

Niccolò Guicciardini

2.1 Foreword

We should not be fooled by Newton's triumph in the Age of Enlightenment. Despite what we might call Newton's seeming apotheosis—as brought about by the public of the Italian Republic of Letters in the mid-eighteenth century—Newtonianism struck those mathematicians, astronomers and physicists who adhered to it as an open system: a field of research filled with unsolved questions. The historical accuracy of this claim may be appreciated by considering the case of Bologna and that of Laura Bassi in particular, where—as Cavazza,[1] Ceranski[2] and Findlen[3] have shown—Newton's work was approached more as a repository of open problems than as a closed doctrine.

2.2 Mathematics

Newton is widely known as a great mathematician. Yet, the mathematical legacy he left his followers was certainly a complex one. The works Newton published at a mature stage in his career were written using a range of different mathematical languages. Besides, Newton presented his theorems and problems with well-known conciseness and incompleteness. Some commentators—such as Jean Paul de Gua de

[1] Marta Cavazza (2006).
[2] Beate Ceranski (1996).
[3] Paula Findlen (1993).

N. Guicciardini (✉)
Department of Philosophy "Piero Martinetti", Milan University, Via Festa del Perdono, 7, Milan, Italy
e-mail: niccolo.guicciardini@unimi.it

L. Cifarelli and R. Simili (eds.), *Laura Bassi–The World's First Woman Professor in Natural Philosophy*, Springer Biographies,
https://doi.org/10.1007/978-3-030-53962-7_2

Malves with reference to the enumeration of cubics featured in the appendix of the 1704 edition of the *Opticks*—attributed these gaps to the genius of a mathematician who all too often soared at heights unreachable by mere mortals.

> Ce géomètre, dont tous les ouvrages portent un caractère singulier de sublimité, paroît en particulier dans celui-ci s'être élevé à une hauteur immense, à laquelle tout autre génie moins pénétrant, et moins fort que le sien, auroit tenté vainement d'atteindre: mais la route qu'il a tenuë dans une entreprise si difficile, se dérobe aux yeux de ceux qui apperçoivent avec étonnement le degré d'élévation auquel il est parvenu. On doit cependant en excepter quelques legeres traces qu'il a eu soin de laisser sur son passage, aux endroits qui avoient mérité qu'il s'y arrêtât plus long-tems. Ces endroits, au reste, sont presque toujours assez distants les uns des autres. Si l'on se propose donc de suivre la même carrière, on est obligé se de guider soi-même dans de longs intervales.[4]

The two mathematicians Vincenzo Riccati and Girolamo Saladini, based in Bologna, felt irritation rather than admiration for the elliptical style of the *Enumeratio*.

> Isaacus Newtonus enumerationem linearum tertii gradus in lucem protulit, licet nulla edita demonstratione, regulisque quibus usus erat, minime attactis, quippe qui magis sibi ipsi admirationem comparare, quam alios edocere cupiebat.[5]

For a variety of reasons which cannot be explored in the present article, in his mathematical works Newton left the—often far from straightforward—task of completing his demonstrations up to his readers.

This is evidently the case in the *Principia*. Here readers must often refer to properties of conics which require knowledge that is not readily available. Other propositions take for granted an acquaintance with techniques for the squaring of curves which left even mathematicians of the likes of Gottfried Wilhelm Leibniz and Christiaan Huygens at a loss. Newton, for instance, starts many of the propositions in the *Principia* (such as proposition 41, Book 1) with the premise that the demonstration implies a method for the squaring of curves (*concessis curvilinearum figurarum quadraturis*); he reduces the problem examined to the squaring of a curve and then—usually in a corollary—presents a solution which depends on this squaring, yet without ever providing any details on how to perform the operation itself.[6]

It was precisely the piecemeal quality of the demonstrations of the *Principia* and the mathematical appendices of the *Opticks*, *Enumeratio linearum tertii ordinis* and *De quadratura curvarum* which prodded early seventeenth-century mathematicians to comment upon, complete and often bitterly criticise Newton's mathematical work. In the context of the polemic between Newton and Leibniz, the supporters of the German mathematician attributed these gaps to Newton's incompetence in the field of infinitesimal calculus: his elliptic style—so they argued—betrays the inferiority of his method compared to that of Leibniz. For these and other reasons, the *Principia* long remained a puzzling text for those seeking to embark on a mathematically informed reading of it. As late as 1716, Eustachio Manfredi wrote that the language

[4] Jean Paul de Gua de Malves (1740).

[5] Vincenzo Riccati, Girolamo Saladini (1765–67).

[6] Bruce Brackenridge (2003).

of Newton's *magnum opus* was no more comprehensible to him than Arabic, and wished that Giuseppe Verzaglia might complete the commentary of the *Principia* he had promised. Nor should we attribute the difficulty faced by Manfredi to his provincialism as a Bolognese: for much the same bafflement was expressed in the Hague, Paris and Basel.[7]

It should further be noted that the very mathematical methods of Newton considerably vary from work to work. Take the collection of mathematical writings edited by William Jones in 1711 (a copy is now to be found in the University Library of Bologna).[8] We here find early works such as *De analysi per aequationes numero terminorum infinitas*, which makes use of infinitesimals (or *momenta*, according to Newton's terminology), alongside more mature works such as *De quadratura curvarum*, where Newton instead claims he only wishes to employ finite magnitudes and passages to the limit. Finally, in his *Principia* Newton adopts a geometrical language and seldom makes use of infinite series or symbolic procedures. The work begins with a section on the "method of first and last ratios", which presents a geometrical theory for the passages to the limit necessary to define the area of a curvilinear surface, and the tangent and radius of curvature of a plane curve. Anyone approaching Newton's mathematics was bound to ask himself what method should be used and what course followed to carry on his mathematical legacy. Was one to employ geometry or algebra? Infinitesimals or limits? Such questions were not raised on a purely technical-mathematical level, since they were also bound to affect one's stance in heated debates such as the one focusing on the contrast between the geometry of the *Veteres* and the algebra of the *Recentes*, between Leibniz's infinitesimals and Newton's "evanescent quantities". As Luigi Pepe has shown in his studies on Italian mathematical treatises, Bolognese, and more generally Italian, mathematicians chose to adopt a Newtonian approach based on limits without thereby renouncing Leibniz's more flexible differential notation.[9,10] Bassi presents this as quite a natural choice in *De problemate quodam mechanico*, where she employs Newtonian terminology for variable magnitudes—termed *fluentes*—while adopting Leibniz's differential notation for her simple calculations.[11] In this regard, mention must also be made of Gabriele Manfredi's research on differential equations, and of the support of Leibnizian methods voiced by Verzaglia, who had studied under Johann Bernoulli in Basel.

[7]Cesare S. Maffioli (1994).

[8]Isaac Newton (1711).

[9]Luigi Pepe (1988)

[10]Luigi Pepe (1984).

[11]Laura Bassi (1757).

2.3 The Principia

The *Principia* is no doubt best regarded as a work which mathematises natural philosophy to an astoundingly sweeping and detailed degree. Before 1687, mathematics had been successfully applied to statics, collisions, parabolic projectile motion and the motion of the cycloid pendulum (Huygens' *Horologium* dates from 1673); Huygens had also made some progress in the study of the motion of projectiles through resisting media (his underlying hypothesis being that resistance is proportional to speed). In the *Principia*, Newton ventured to examine phenomena of a level of complexity quite unforeseen by his predecessors: take his (qualitative) study of the motion of three gravitationally interacting bodies in Sect. 11 of Book 1, of the attraction of extended bodies in Sects. 12 and 13 of Book 1, or of the motion of projectiles through resisting media—under the hypothesis that resistance to motion varies as the linear combination between a term proportional to the speed and one proportional to the square of the speed—in Sects. 3 and 4 of Book 2. In Book 3, planetary astronomy is mathematised, ultimately lending physical-mathematical confirmation to the Copernican system.[12]

As is widely known, the whole Copernican issue was still being made the object of censures which could hardly be overlooked in Bologna. Experimental confirmation of the motion of the Earth was later found in the phenomenon of the aberration of light, which was recorded and interpreted by James Bradley; in 1729 Eustachio Manfredi published his own observations on the phenomenon, although when doing so he was first forced to state his own personal adherence to geocentricism.[13] In the third book of the *Principia*, Newton presents some striking results concerning the ebb and flow of tides, the procession of the equinoxes, the orbit of comets, the shape of the Earth and the perturbations of planetary orbits, while also broaching the difficult problem of the motion of the Moon. These are all undeniable achievements. Still, the *Principia* initially struck competent readers as a rather puzzling text. The secondary literature on the reception of Newton's work has often focused on the issue of the nature of gravitation. Many readers with a Cartesian background were perplexed by this force acting from a distance, for which no mechanical explanation was provided. The debate which ensued is well known, but it is not on this great riddle which I wish to focus. Rather, I would like to stress how many astronomers, mathematicians and natural philosophers found an array of unsolved problems in the *Principia*, and how the reception of Newtonianism also came about through an acceptance of Newton's challenge to solve what he himself had only solved *in nuce*—or approached from a wrong angle. I will here list some of the questions Newton left open. Each of these matters provided a stimulus for eighteenth-century mathematicians to develop an approach that might prove more satisfying from a physical as well as mathematical point of view.

First of all, it must be observed that Newton's treatment of the motion of fluids in the second book of the *Principia* could hardly be used by researchers interested

---

[12] I. Bernard Cohen (1999).

[13] Eustachio Manfredi (1729).

in either ballistics or water management. Engineering water management studies were indeed made by mathematicians associated with Bologna. However, only with the completely new approach to hydrodynamics developed by Johann Bernoulli, his son Daniel and Leonhard Euler, among others, were initial steps taken in the right direction. New mathematical tools, such as partial differential equations, unknown to Newton, enabled this new chapter in the history of analytical mechanics to be written. In this regard, mention must be made of Bassi's *De problemate quodam hydrometrico*, a work devoted to the outflow of liquid from a container with an opening made at its bottom.[14] Newton had already examined this problem in the second book of the *Principia*. In fact, he had been forced to completely rewrite this section for his second edition of the work (1713). The idea, which had first been suggested by Evangelista Torricelli and refined by Edme Mariotte, Domenico Guglielmini and Huygens, among others, consisted in working out the speed of the outpouring fluid in function of its height.[15]

In Newton's day, the shape of the Earth was not quite clear. According to the Cartesians, the Earth was shaped like a "melon": they believed that the distance between the North and South Pole had to be greater than the diameter of the equator. Newton, by contrast, believed that the Earth was flattened at its poles: a theoretical estimate which was confirmed in the eighteenth century through the expeditions that measured meridian arcs in Lapland and Peru. Newton reasoned as follows. He envisaged the Earth as a homogeneous and rotating fluid mass. The shape of the Earth, he argued, must be the equilibrium shape of this mass. If the mass is in equilibrium, the following must hold true: if we imagine solidifying the entire mass with the exception of two mutually communicating rectilinear channels—the first linking the North Pole to the centre of the Earth, the second connecting a point at the equator with the centre of the Earth—the fluid of the two channels will be found to be in equilibrium. Their length must therefore be such that the fluid they contain will remain still. Because of the centrifugal force generated by the rotation of the Earth, it will only be possible to achieve this equilibrium if the equatorial channel is longer than the polar one. By taking account of the speed of rotation of the Earth, of the mathematical results (proposition 91, Book 1) for the attraction exercised by an ellipsoid of revolution on a mass point situated on the extension of its axis, and of empirical data regarding the variation of the oscillation period of a pendulum as a function of latitude, Newton obtained an approximate measurement for the Earth's flattening at its poles. Note that Newton did not demonstrate what the surface shape of a rotating fluid mass must necessarily be. Yet, his principle of "solidification" has played an important role in the study of the equilibrium of fluids. Newton also carefully examined the variation of gravity according to latitude. It must be observed, as George Smith has often emphasised, that this is the only result in the *Principia* crucially dependent upon the universality of the Law of Gravitation, which is to say upon the fact that gravity must apply to each constitutive particle of the Earth—and not just to heavenly bodies, at a macroscopic level. These geodesic results are

[14]Laura Bassi (1757).

[15]George Smith (2001).

obtained by assuming that the density of the Earth is homogeneous; hence, they are only valid in a very approximate way.[16]

Newton reached some interesting results concerning the motion of the Moon. He succeeded in lending satisfying mathematical expression to some anomalies in the planet's motion: its divergence from the Law of Areas, the motion of its line of nodes, the fluctuation of the inclination of the plane of its orbit. As previously noted, the Sun generates forces which affect the Earth-Moon system, making the motion of our satellite irregular. One of these anomalies, the precession of the lunar apogee, escaped Newton's analysis and became a celebrated problem addressed by all the leading mathematicians of the eighteenth century, from Euler to Laplace.[17]

The Copernican System attributes the procession of the equinoxes to a gradual shift in the rotation axis of the Earth. This axis is not fixed in relation to the stars: while keeping a fixed tilt in relation to the terrestrial orbit, it traces a cone. The complete precession period is of about 26,000 years. Newton's explanation is as follows. Since the Moon's period of revolution around the Earth is very small compared to the precession of the terrestrial axis, we can envisage the mass of the Moon as a ring distributed around the Earth. The rotation axis of the Earth is tilted in relation to the plane of this ring. The Earth, moreover, is flattened at its poles: it may be conceived of as a sphere with an "equatorial bulge". The result is the subjecting of a non-spherical rotating body to a torque. While Newton never explored the dynamics of rigid bodies, he realised that the effect exercised by the lunar ring on the equatorial bulge of the Earth generates a conical motion of precession of the rotation axis. The same kind of reasoning made for the Moon may be applied to the Sun: for the gravitational action of the Sun must be regarded as one of the causes of the precession of the equinoxes. Finally, it is easy to realise that if the Earth were perfectly spherical, there would be no such thing as the precession of the equinoxes. This was the first physical explanation ever provided for an astronomic phenomenon which had been known since Antiquity. It must nonetheless be observed that Newton did not know the mass of the Moon and hence did not know how to gauge the intensity of the gravitational pull exercised by our satellite upon the equatorial bulge of the Earth. Ultimately, what he did was to assign the lunar mass a value that would enable him to reach the result he expected. Besides, Newton did not possess a theory for the dynamics of rigid bodies and his discussion of the precession of the equinoxes had to be completely revised in the light of Euler's results on the dynamics of extended bodies.[18]

The way Newton inferred the phenomenon of tides from the theory of gravitation was by showing that the pull of the Moon (and, to a lesser extent, the Sun) upon the ocean masses generates two swellings. Consider, for instance, the gravitational pull of the Moon: this will be stronger on the water particles closer to our satellite and weaker on those situated at the antipodes. It is this discrepancy which causes two swellings of the ocean surface. Since the Earth spins on its axis, the two swellings

[16]John L. Greenberg (1995).
[17]Curtis Wilson (2001).
[18]Clifford Truesdell (1960).

will cover the whole terrestrial surface, producing two high tides and two low ones in 24 h. This theory had to be radically revised in the eighteenth century, when scientists discovered that the force at work is not the one perpendicular to the surface of the water, but the one tangent to it and responsible for the horizontal flow of the water particles. Newton's theory, moreover, is a static one, which fails to take account of the fact that tides are a dynamic phenomenon. To put it briefly, according to the theory developed by Laplace, tides are wave-motions produced by periodical gravitational disturbances from the Sun and Moon. These motions are influenced by a wide range of factors, including the Coriolis effect and the geometry of ocean basins, which favours given frequencies in the oscillation of ocean masses.[19]

As noted above, the great value of Newton's results is self-evident. A single force, the familiar gravitational force responsible for the falling of bodies towards the centre of the Earth, also explains a large number of terrestrial and celestial phenomena. This force may be subjected to mathematics, which will then become the language enabling us to grasp the causes behind the most varied phenomena. And it was precisely the methamaticisation of the force of gravity which enabled solutions to be found at the time for the greatest puzzles surrounding the "World System": is there any good reason to choose the Copernican system over the geocentric? And if the planets revolve around the Sun, what is it that keeps them within their orbits?

The solution to these puzzles provided in the *Principia*, while valuable in itself, is beset by difficulties of both a technical and foundational sort. The former I have already referred to above by emphasising how in the second and third book of the *Principia* Newton explores many topics (the motion of fluids, that of the lunar apogee, the shape of the Earth, the tides, the procession of the equinoxes) through physical and mathematical tools which retrospectively strike us as inadequate. It was up to mathematicians such as Clairaut, Daniel Bernoulli, Euler, d'Alembert, Lagrange and Laplace—to mention but a few names—to develop the dynamics of fluids and rigid bodies, the calculus of variations and the theory of partial differential equations, the least action principle, elliptical integrals, the series expansion of trigonometric functions, and many other techniques which enabled the solving of those problems inadequately addressed in the *Principia.* The reception of Newtonianism among mathematicians, astronomers and physicists in the eighteenth century must be envisaged not as the embracing of a world system on their part, but rather as an expression of their desire to take part in a research programme that Newton had left open—and this precisely because many of the problems tackled in the *Principia* required new tools in order to be solved.

2.4 Opticks

The open character of Newton's legacy is particularly evident in the case of *Opticks*, which ends with a series of queries which Newton couches in a hypothetical and

[19]Martin Ekman (1993).

speculative language. These queries ultimately outline a research plan still awaiting to be implemented for the study of chemical, electrical and magnetic phenomena. Their role in shaping enquiries into the so-called "Baconian sciences"—to adopt Thomas Kuhn's terminology—in the eighteenth century can hardly be overemphasised.[20] Laura Bassi, who also made contributions in the fields of mechanics and hydrometry (see notes 11, 14), was especially fascinated by this aspect of Newton's legacy. After all, *Opticks* was a significant point of reference for the Bolognese—at least from 1728, when Newton's experiments on refraction were confirmed by Francesco Maria Zanotti and Francesco Algarotti at the Institute, proving Giovanni Rizzetti's arguments wrong.[21] This successful outcome of the *experimentum crucis* in Bologna, while highly meaningful for the reception of Newton's theories on light, ought not be seen as the straightforward embracing of a theory devoid of any gaps and leaving no questions open.[22]

It is worth noting how Newton's queries regarding chemical and electric phenomena were also grounded in his interest for the phenomena of perception and volition, which is to say vital phenomena that fell within the sphere of interests of the Veratti-Bassi couple. To further appreciate this aspect of the queries, one might turn to the Newtonian manuscript which ultimately laid their foundation, *An Hypothesis Explaining the Properties of Light*. Prodded by the criticism he had received from Robert Hooke and Huygens, Newton presented this text at the Royal Society in December 1675.[23]

In his *Hypothesis*, Newton argues that space is pervaded by very fine aether, a kind of fluid possessing great elasticity. This fluid may be found not just in empty space, but also in bodies, for it permeates the pores of crystals, glass and water. In empty space, however, it occurs in a denser form than in solids. Through this aether vibrations travel that are similar to acoustic vibrations, only much faster and more minute (nowadays one would say they have a much shorter wavelength). Light, according to Newton, consists in a flow of corpuscles of various form which interact with the aether: aether refracts light and light warms aether. When these corpuscles meet the surface of a reflecting or refracting medium, they generate aether waves, like stones falling into a pond. According to this hypothesis, light corpuscles all move at the same speed. Aether refracts light: the corpuscles, that is, tend to deflect towards regions where the aether is less dense; and since aether is less dense in glass than in air, the luminous corpuscles are deflected when they pass through the surface of separation between air and glass. When they pass from air to glass—Newton explains—they are accelerated in the normal direction of the surface. Newton, therefore, used this model to explain the mechanism behind refraction.

As has just been mentioned, when the corpuscles meet the surface separating two media with different refractive indexes, they set the aether in vibration, and

[20]Thomas Kuhn (1977).

[21]Marta Cavazza (2002).

[22]Paolo Casini (1993).

[23]Isaac Newton, an hypothesis explaining the properties of light discoursed in my severall papers, in H.W. Turnbull (1959).

this in turn lends periodic properties to light. The phe-nomena Hooke observed on thin surfaces (soap bubbles, mica, etc.) and recorded in his *Micrographia* may be explained, according to Newton, by attributing periodic properties to light—something which supporters of the wave theory would have found much easier to do.

Confirmation of the fact that the aether invoked by Newton is a vital element comes from many passages of the *Hypothesis*, and particularly those concerning the relations between volition and muscular movement. Newton refers to a "puzzling" problem: how are muscles contracted and dilated in such a way as to generate movement in animals? The most popular answer was that provided by the Cartesians, who believed the nervous system to consist of thin channels through which the animal spirit flows, this being a gaseous or possibly fiery substance filling the nerves of the body and the pores of the brain. Volition and perceptions were thus explained by the Cartesians by positing a hydraulic exchange between the brain, from which the animal fluid was believed to flow, and the limbs of the body, which were held to be driven by this fluid when it steeped into the muscles. The hypothesis suggested by Newton is not far from the Cartesian: the aether permeating muscles he regarded as an animal spirit capable of dilation and contraction. According to his view, the soul has the power to fill muscles with this spirit or "wind" through the nerves. It is not necessary—Newton adds—to posit a very large variation in the density of the aether in muscles, since thanks to its considerable elasticity all it would take would be a small variation in density to engender a big variation in pressure. Newton's language, however, is more reminiscent of Henry More and Thomas Willis than it is of Descartes.

In the closing queries of *Opticks*, Newton reframes his 1675 hypothesis on aether according to the principles of attraction and repulsion. It is nonetheless clear that he is still approaching themes such as those of the elasticity of air, the short-range forces behind chemical affinities and electric forces starting from an idea of matter that cannot be reduced to the "inert and brute" matter of Cartesian mechanics. Newton's research on elastic fluids—such as air and the electric fluid—presupposes a conception of matter as shaped by active principles; this, in turn, led many Newtonians who carried on the programme outlined in Newton's *Opticks* to cultivate an interest in the world of life.[24] Indeed, Bologna was destined to become an important centre in the debate on the therapeutic applications of electricity, a debate which peaked with Luigi Galvani's work in the late eighteenth century.

---

[24] Maurizio Mamiani, Emanuela Trucco (1991).

2.5 Closing Remarks

Laura Bassi's physics would appear to have been influenced chiefly by *Opticks*, and especially its queries, although she also made a number of contributions to mathematical physics inspired by the *Principia.* Newton's queries, as is well known, are literally open questions which the mathematician posed his followers, as if seeking to suggest the possible lines of research they ought to follow. What I wished to draw attention to in this paper was the fact that the open character of Newton's heritage also extends to his mathematical work and the *Principia.* Italian Newtonians of the eighteenth century, in whose ranks Laura Bassi may certainly be counted, often regarded Newton's work not as a system one might convert to, but rather as a repository of open problems on which to focus one's research.

Figures 2.1, 2.2, and 2.3 show a portrait of Isaac Newton and other related images.

Fig. 2.1 Isaac Newton (1642–1727), Roman school, oil on canvas, 18th century, Biblioteca Universitaria di Bologna, inv. 466

Fig. 2.2 Giambattista Pittoni the younger, Domenico and Giuseppe Valeriani, *An allegorical Monument to Sir Isaac Newton*, 1727–29, oil on canvas, Cambridge, The Fitzwilliam Museum; Museum accession number PD.52-1973. The beam of light refracted by a prism showed the spectrum of colours comprised by sunlight

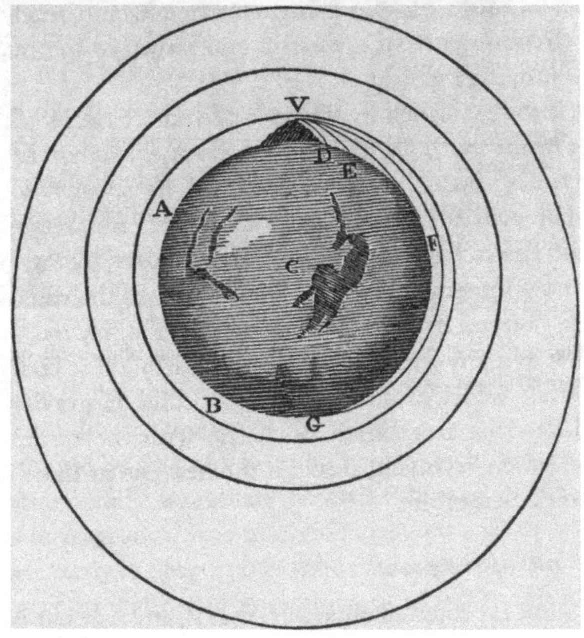

Fig. 2.3 In this famous picture Newton shows the universality of the gravitational force (I. Newton, A Treatise of the System of the World, 2nd ed., Fayram, London 1731

Acknowledgements This research was funded by the Department of Philosophy "Piero Martinetti of the University of Milan under the Project "Departments of Excellence 2018-2022" awarded by the Ministry of Education, University and Research (MIUR).

References

Laura Bassi, *De problemate quodam mechanico*, in *De bononiensi scientiarum et artium Instituto atque Academia commentarii*, IV, 4, 1757, pp. 74–79.

Laura Bassi, *De problemate quodam hydrometrico*, in *De bononiensi scientiarum et artium Instituto atque Academia commentarii*, IV, 1757, pp. 61–73.

Bruce Brackenridge, *Newtons' Easy Quadratures: "Omitted for the Sake of Brevity"*, «Archive for History of Exact Sciences», LVII, 2003, pp. 313–36.

Paolo Casini, *The reception of Newtons' Opticks in Italy*, in *Renaissance and Revolution. Humanists, Scholars, Craftsmen and Natural Philosophers in Early Modern Europe*, edited by Judith V. Field and Frank A.J.L. James, Cambridge University Press, Cambridge 1993, pp. 215–227.

Marta Cavazza, *The Institute of Science in Bologna and the Royal Society in the Eighteenth Century*, «Notes and Records of the Royal Society», LVI, 1, 2002, pp. 3–25.

Marta Cavazza, *Una donna nella repubblica degli scienziati: Laura Bassi e i suoi colleghi*, in *Scienza a due voci*, edited by Raffaella Simili, Olschki, Firenze 2006, pp. 61–85.

Beate Ceranski, *"Und sie fürchtet sich vor niemandem": Die Physikerin Laura Bassi (1711–1778)*, Campus Verlag, Frankfurt/NewYork 1996.

I. Bernard Cohen, *A Guide to Newtons' Principia*, in Isaac Newton, *The Principia: Mathematical Principles of Natural Philosophy*, translated by I. Bernard Cohen and Anne Whitman, with the assistance of Julia Budenz, University of California Press, Berkeley 1999, pp. 1–370.

Jean Paul de Gua de Malves, *Usages de l'Analyse de Descartes. Pour Découvrir, Sans le Secours du Calcul Differentiel, les Proprietés, ou Affections Principales des Lignes Géométriques de Tous les Ordres*, Chez Briasson et Piget, Paris 1740, pp. xi–xii.

Martin Ekman, *A concise history of the theories of tides, precession-nutation and polar motion (from antiquity to 1950)*, «Surveys in Geophysics», XIV, 1993, pp. 585–617.

Paula Findlen, *Science as a Career in Enlightenment Italy: The Strategies of Laura Bassi*, «Isis», LXXXIV, 3, 1993, pp. 441–469.

John L. Greenberg, *The problem of the Earths' shape from Newton to Clairaut. The rise of mathematical science in eighteenth-century Paris and the fall of "normal science"*, Cambridge University Press, New York 1995.

Thomas Kuhn, *Mathematical versus experimental traditions in the development of physical science*, in *The Essential Tension*, University of Chicago Press, Chicago 1977, pp. 31–65 (Italian translation, Einaudi, Torino 1985). Previously published in «Journal of Interdisciplinary History», VII, 1, 1976, pp. 1–31.

Cesare S. Maffioli, *Out of Galileo: the science of waters 1628–1718*, Erasmus Publishing, Rotterdam 1994, p. 21.

Maurizio Mamiani, Emanuela Trucco, *Newton e i fenomeni della vita*, «Nuncius», VI, 1991, pp. 69–96.

Eustachio Manfredi, *De annuis inerrantium stellarum aberrationibus*, Typis Constantini Pisarri, Bologna 1729.

Isaac Newton, *Analysis per Quantitatum Series, Fluxiones, ac Differentias. Cum Enumeratione Linearum Tertii Ordinis*, edited by William Jones, ex Officina Pearsoniama, London 1711.

Luigi Pepe, *Sulla trattatistica del Calcolo infinitesimale in Italia nel secolo XVIII*, in Atti del Convegno "La storia delle matematiche in Italia" (Cagliari, 1982), Tip. Monograf, Bologna 1984, pp. 145–227.

Luigi Pepe, *Newton, il metodo delle flussioni e i fondamenti dell'analisi in Italia nel secolo XVIII*, in Atti del Convegno "Storia degli studi sui fondamenti della matematica", Tip. Luciani, Roma 1988, pp. 185–224.

Vincenzo Riccati, Girolamo Saladini, *Institutiones Analyticae*, Ex Typographia Sancti Thomae Aquinatis, Bologna 1765–67, p. x.

George Smith, *The Newtonian style in Book II of the Principia*, in *Isaac Newtons' Natural Philosophy*, edited by Jed Z. Buchwald and I. Bernard Cohen, MIT Press, Cambridge (Mass) 2001, pp. 249–313.

Clifford Truesdell, *A program toward rediscovering the rational mechanics of the Age of Reason*, «Archive for History of Exact Sciences», I, 1960, pp. 3–36.

H. W. Turnbull (Editor), *The Correspondence of Isaac Newton*, vol. 1, Cambridge University Press, Cambridge 1959, pp. 362–386.

Curtis Wilson, *Newton on the Moons' Variation and Apsidal Motion. The Need for a Newer "New Analysis"*, in *Isaac Newtons' Natural Philosophy*, edited by Jed Z. Buchwald and I. Bernard Cohen, MIT Press, Cambridge (Mass) 2001, pp. 139–88.

Chapter 3
Physics in the Eighteenth Century: New Lectures, Entertainment and Wonder

Sofia Talas

> Pénétré de respect, & même de reconnoissance pour les grands-hommes qui nous ont fait part de leurs pensées, & qui nous ont enrichis de leurs découvertes, de quelque Nation qu'ils soient, & dans quelque tems qu'ils aient vécu, j'admire leur génie jusques dans leurs erreurs, & je me fais un devoir de leur rendre l'honneur qui leur est dû; mais je n'admets rien sur leur parole, s'il n'est frappé au coin de l'expérience. En matière de Physique, on ne doit point être esclave de l'autorité; on devroit l'être encore moins de ses propres préjugés, reconnoître la vérité par-tout où elle se montre [...][1]

Thus wrote in 1743 Jean-Antoine Nollet, professor of experimental physics at the *Collège de Navarre*, scientist of the French court and *Maitre de Physique* of the king's children. He also explained in 1770:

> Songez que s'il vous est permis de fixer l'attention de vos Auditeurs par des phénomènes qui les surprennent, il n'est pas de la dignité d'un Physicien de leur laisser ignorer les causes, quand il peut les leur faire connoître; ainsi, quoique le verre soit fragile, il faut le faire entrer dans la construction des machines de Physique préférablement au métal & aux autres matières opaques, toutes les fois qu'on pourra s'aider de sa transparence pour faire voir le méchanisme des opérations: car je le répète, notre premier point de vue doit être d'enseigner, d'éclairer, & non de surprendre ou d'embarrasser.[2]

---

[1]Nollet (1743–1764, vol.1, 1743, pp. xx–xxi) ("Full of respect and gratitude towards those great men who communicated us their thoughts and enriched us with their discoveries, whatever country they were from and whenever they lived, I admire their sublime and deep minds, even their mistakes. I give them that honour that I owe them and that they deserve; but I accept neither what they say, nor anything that cannot be proved by experiments. As for Physics, we cannot be slaves of the authority, and we ought not be subjected to our own prejudices, but we have to meet and receive the truth, wherever it appears").

[2]Nollet (1770, vol. 1, p. xx) ("Think that though you are allowed to catch the attention of your audience through phenomena which surprise them, letting them ignore the causes, when these can be known, does not fit the dignity of a physicist. So, though glass is fragile, it is necessary to introduce it in the construction of Physics machines, preferring it to metal or to other opaque materials, whenever transparency could help us to see the mechanism of processes, because, as I already said, our first aim must be to teach, enlighten, and not to surprise or embarrass").

S. Talas (✉)
Museum of the History of Physics, Padua University, Padova, Italy
e-mail: sofia.talas@unipd.it

L. Cifarelli and R. Simili (eds.), *Laura Bassi–The World's First Woman Professor in Natural Philosophy*, Springer Biographies,
https://doi.org/10.1007/978-3-030-53962-7_3

Fig. 3.1 A lecture of experimental physics in the eighteenth century (Jean-Antoine Nollet, *Leçons de physique expérimentale*, 6 vol., Guérin, Paris 1743–1764)

The very essence of physics practice in the Age of Enlightenment emerges from these words. At that time, both physics research and teaching were solidly based on observations and experiments. The latter had already become crucial elements of research through the Scientific Revolution, but it is in the course of the eighteenth century that the physics lectures outlined by Nollet, based on demonstrations—often spectacular ones—, asserted themselves (Fig. 3.1). For the purpose of this new way of lecturing, new specific devices were designed, and more and more collections of scientific instruments—the so called "Cabinets of Physics"—were set up. These were to become in several cases the starting cores of nineteenth-century laboratories. In this paper, after outlining how the new lectures spread throughout Europe, we will examine their main features and the extraordinary popularity they acquired not only among specialists, in academies or universities, but in *salons* and royal courts as well. Science had turned from private to public.

Let us recall at first that until the Middle Ages, in Europe, knowledge of Nature depended on direct observation, there were very few instruments and experiments were limited. The science of nature, or natural philosophy, mainly illustrated and glossed the knowledge contained in the works of the "old masters", whose authority was hardly ever called into question. In the fifteenth and sixteenth centuries, the time of Humanism and Renaissance, a new cultural ferment led to rediscover and study Greek and Muslim science, man recovered confidence in his own potentialities and manual arts started being regarded as a way of producing knowledge. Mathematics-based instruments were more and more made and used. Through such devices, mathematics—geometry in particular—was applied to different "arts", such as topography, navigation or astronomy, reforming them in the same way as the science of perspective was reforming at that time the art of painting. All these changes laid the foundations for the great revolution which was to be started by Galileo Galilei.

Galilei radically transformed the approach to scientific research by placing at the centre of his investigations experiments and observations which alone, according to him, could support or refute hypotheses and ideas, without being subject to the authority of the knowledge contained in books. He regarded instruments as crucial: objects created by manual arts—the telescope in primis—can produce knowledge. This marks the symbolical starting point of that interweaving of science and technology which was to be one of the features of modern science. Along with experiments and observation, Galilei praised accurate mathematical demonstrations and he thought that natural phenomena could be translated into precise mathematical laws. Galilei thus opposed the distinction that was made at the time between mathematicians—who were expected to carry out measures and calculations—and natural philosophers—the only ones who were supposed to study and know nature. It is not surprising at all that Galilei, who had been professor of mathematics during his staying in Padua, asked the Gran Duke of Tuscany to appoint him "mathematician and philosopher", when he moved to Florence in 1610 (Fig. 3.2).

It is well-known that Galileo's way of practicing science was severely criticised and the new experimental method, which owes so much to him, asserted itself very slowly in the course of the seventeenth century. In the 1660s, for instance, Robert Boyle was attacked by Thomas Hobbes due to his experiments in rarefied air. Hobbes criticised Boyle for his using an air-pump which, according to Hobbes, raised spectacular phenomena, but without any kind of connection with the real knowledge of nature.

Learned societies and academies that were spreading all over Europe in those years—the *Accademia del Cimento*, for instance, founded in Florence in 1657 by the Prince Leopoldo De' Medici and the Grand Duke Ferdinando II, or the *Académie des Sciences*, created in 1666 by Louis XIV—were the breeding grounds where experimental natural philosophy, the new scientific practice, developed with more strength. Experiments and demonstrations about various aspects of the natural world—properties of air, minerals, plants or animals were, for instance, analysed—were regularly

Fig. 3.2 Portrait of Galilei, philosopher and mathematician of the Grand Duke of Tuscany (Galileo Galilei, *Il Saggiatore*, G. Mascardi, Roma 1623)

held, attended by the academies' members. Only in the last years of the seventeenth century, some courses illustrated by experiments started being introduced in universities. From 1695, for example, Pierre Polinière held lectures of what he called "experimental physics", which were included in the philosophy courses of the University of Paris. The word "physics" was meant at that time in a broad sense and was a synonym of "natural philosophy". Polinière's lectures, gathered in a booklet whose title was *Expériences de Physique*, published in 1709, included experiments on hydrostatics, pneumatics, acoustics, magnetism and chemistry as well as experiments on the anatomy of animals and plants.[3] Stimulated by Boyle's works on pneumatics, Burchard de Volder also started giving lectures of experimental physics in Leiden from 1675, but these were only sporadic undertakings compared to what was to happen.

At the beginning of the eighteenth century, John Keill and William Whiston, both Newton's pupils, introduced new experimental lectures in Cambridge and Oxford

[3]Polinière (1709).

and experiments-based lectures were proposed in London to a wide public. The connection with the Royal Society, chaired by Newton from 1703, was tight. These lecture-demonstrations widely drew on what had been performed during previous meetings of the Royal Society, but presented only experiments on mechanics, hydrostatics, pneumatics, heat and optics. They were defined as "experimental physics" or "experimental philosophy" lectures. Most of the lecturers were or became *fellows* of the Royal Society, like James Hodgson, for instance, who was the first to give a public physics course in 1705, or Francis Hauksbee, who was also an instrument-maker and curator of experiments at the Royal Society. Hauksbee actively contributed to Hodgson's lecture-demonstrations and succeeded Hodgson as a lecturer from 1709 onwards.[4]

The resources provided by the scientific instruments market, which was flowering in London at that time, were crucial for the success of the new lectures. In fact, the Scientific Revolution had brought the development of professional scientific instrument makers, who included in their production recently invented devices, like telescopes, microscopes, barometers, thermometers, clocks or air-pumps. Such makers were pretty rare on the Continent, but widespread in the English capital, where they could enjoy a wide range of customers, from mariners to private citizens, more and more eager to own instruments such as clocks and barometers. These craftsmen became precious collaborators for the lecturers and made specific instruments for the new experimental physics lectures.

One of the most prominent lecturers of the time was John Theophilus Desaguliers (1683–1744), who attended John Keill's experimental philosophy course at the University of Oxford. He replaced Keill for some years, then moved to London, where he gave public physics lectures from 1713. He succeeded Hauksbee at the Royal Society both as a *fellow* and as curator of experiments. He travelled a lot in England and abroad, contributing significantly to the success of the new public physics lectures. He collected his lectures in a book *A course of experimental philosophy*, which was translated into many different languages and was very popular during the whole eighteenth century.[5]

Two Dutchmen, Willem Jacob 's Gravesande (1688–1742) and Pieter van Musschenbroek (1691–1761), then played a central role in spreading the new physics lectures on the Continent.[6] 's Gravesande, who studied as a lawyer but whose real interest lay in science, went in 1715 to England, where he met Newton, becoming a strenuous supporter of his ideas. In 1717 he was assigned the chair of mathematics and astronomy at the University of Leiden and focused on disseminating Newtonian philosophy, which was regarded as brilliant but very difficult to understand. Thanks to the collaboration of Jan van Musschenbroek, instrument maker in Leiden, 's Gravesande devised and had made a wide range of new instruments. In 1718 he wrote to Newton:

[4]Morton and Wess (1993).

[5]Desaguliers (1734–1744).

[6]De Clercq (1997).

Fig. 3.3 Instrument devised by Willem 's Gravesande to demonstrate the parabolic motion of projectiles, mid-eighteenth century, Giovanni Poleni's Cabinet of Physics, Museum of the History of Physics, University of Padua

> I begin to hope that the way of philosophizing that one finds in this book [Newton's *Opticks*] will be more and more followed in this country, at least I flatter myself that I have had some success in giving a taste of your philosophy in this university. As I talk to people who have made very little progress in mathematics I have been obliged to have several machines constructed to convey the force of propositions whose demonstrations they had not understood. By experiment I give a direct proof of the nature of compounded motions, oblique collisions, and the effect of oblique forces and the principal propositions respecting central forces.[7]

Experiments and instruments were thus needed to give direct and immediate demonstrations of the laws of physics. For instance, in order to demonstrate that projectiles follow a parabolic trajectory—as Galilei had discovered in the previous century—'s Gravesande invented a device where a little ball was released along a track, acquiring a horizontal velocity at the end of it (Fig. 3.3). A parabola with its coordinates was drawn on an adjacent vertical panel made of wood and a few rings were fixed along the parabola. The little ball passed through all the rings, thus showing that the composition of the horizontal uniform motion with the vertical uniformly accelerated motion due to gravity led to a parabolic motion. This instrument, as well as many others designed by the Dutch scholar, was to become a classic element within Cabinets of Physics during the whole eighteenth century.

Direct and immediate demonstrations were thus central, but 's Gravesande also presented accurate measurements and detailed analyses of the laws of physics. Furthermore, from the second edition of his well-known treatise *Physices elementa mathematica, experimentis confirmata, sive introduction ad philosophiam Newtonianam*,[8] the scholar systematically put side by side experiments, instruments and

[7]De Clercq (1997, *At the sign...*, p. 76).

[8]'s Gravesande (1720–1721, 1725, 1742).

mathematical demonstrations. Directly influenced by Newton,[9] this work dealt with the knowledge of nature as a branch of mathematics, as the author specified in the preface: "Natural Philosophy is placed among those parts of Mathematics, whose Object is Quantity in general".[10]

The text was published for the first time in 1720–1721 and then extended in the 1725 and 1742 editions. It obtained a great success all over Europe and became one of the fundamental treatises of eighteenth-century physics. It was translated by Desaguliers and printed in English by 1720–21 with the title *Mathematical Elements of Natural Philosophy, Confirm'd by Experiments: or, an Introduction to Sir Isaac Newtons' Philosophy*. It was published in French in 1747. It included sections on the motion of bodies, forces, the study of fluids (hydrostatics and hydrodynamics), pneumatics, "fire"—a section that included the study of heat, light and electricity acquired by friction—, and the description of the Solar System and its motions (Fig. 3.4).

As for Pieter van Musschenbroek, brother of the instrument maker Jan van Musschenbroek, he taught in Duisburg and Utrecht at first, then succeeded 's Gravesande in Leiden. He was an enthusiastic Newtonian too and a friend of 's Gravesande's. He met Desaguliers during a journey in England. His lectures of experimental physics became widely known thanks to his treatises, in particular *Elementa physicae* (Leiden 1734) and the posthumous text in two volumes *Introductio ad philosophiam naturalem* (Leiden 1762), which were translated into Dutch, English, French and German.[11] Very interested in the study of electricity, quite a blossoming topic at the time, as we will see, Pieter van Musschenbroek inserted more and more of it in his treatises.

Excellent lecturers of the new physics, 's Gravesande and van Musschenbroek attracted students from all over Europe. Among them, Jean-Antoine Nollet (1700–1770) was to become one of the main supporters of the new way of teaching.[12] Born in a peasant family, Nollet studied theology at first and became a deacon—he is known as *l'abbé Nollet*. He was very interested in sciences and got in touch with some of the major exponents of French scientific life, in particular Charles François Dufay and René Antoine Ferchault de Réaumur, from whom Nollet learnt elements of physics and laboratory techniques. Dufay and Réaumur themselves took the *abbé* to the Netherlands and to England, where he discovered with enthusiasm the new lectures of experimental physics. Back to France in 1735, Nollet took over Pierre Polinière's courses, introducing the contents and the methodology that had struck him, but with further refinements.

According to the most widespread prejudices of the time, the knowledge of physics was reserved to scholars only. As Nollet wrote, for many people, "la Physique ne

---

[9]The title itself of 's Gravesande's treatise is very close to Newton's *Philosophiae Naturalis Principia Mathematica*, published in 1687, which was translated into English with the title *Mathematical Principles of Natural Philosophy*. It is worth noting that 's Gravesande's Latin title used the word "physics" and the English title used the words "natural philosophy" with the same meaning.

[10]'s Gravesande (1720–1721, p. ii).

[11]Van Musschenbroek (1734, 1762).

[12]Heilbron (1970–1980).

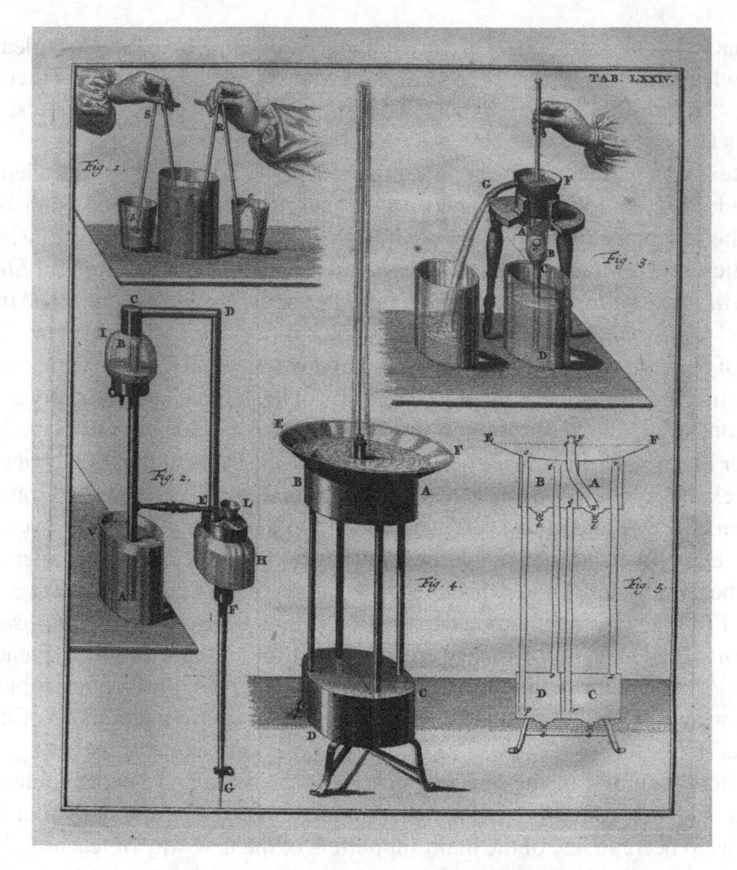

Fig. 3.4 Plate from Willem 's Gravesande *Physices Elementa Mathematica* (1742 edition), illustrating some of the experiments that were proposed to show the action of air pressure on liquids

se présente [...] qu'avec des caractères géométriques & toute hérissée d'Algèbre"[13]. On the contrary, like the Dutch and English lecturers, Nollet wanted to bring physics within the reach of the largest possible number of people. He addressed his own lectures to the most diverse public, not only to specialists and university students who could follow him into details and deeper analyses, but above all to "amateurs", so that physics "devînt un bien dont la possession fût commune à tout le monde".[14] Physics thus entered French *Salons*. The point was to find a delicate balance between the seriousness of the course and the spectacular features of the experiments, combining utility with pleasure ("l'agréable à l'utile").[15] The so-called "mechanical paradoxes",

[13]Nollet (1738, p. xiii) ("Physics appears only [...] with geometrical characters and fraught with Algebra").

[14]Nollet (1738, pp. vi–vii) ("could become a common good owned by everyone").

[15]Nollet (1738, p. xi).

Fig. 3.5 Tantalus vases, mid-eighteenth century, Giovanni Poleni's Cabinet of Physics, Museum of the History of Physics, University of Padua

mentioned in most treatises of the time, are emblematic in this sense. The double cone, for instance, which seemed to go up along a track, surprised and struck the audience, defying common sense, but the lecturers explained to their public that the centre of mass of the double cone was actually going down, as expected from the laws of mechanics. A lecture could also start with "Tantalus vases", recipients made of glass where a fluid was poured. The vases were filled in normally at first but, when they were almost full, the fluid started coming out from underneath spouts. Why? The lecturers then explained that a siphon was fixed at the centre of the vases, they analyzed the way siphons work, and examined also the action of the atmospheric pressure (Fig. 3.5).

Boasting of having defeated several prejudices, Nollet was glad to host women in his audience, specifying that: "On ne croit plus qu'il y ait à rougir de sçavoir ce qu'on pourroit absolument ignorer; on sçait qu'un esprit éclairé n'est point incompatibile avec d'autres agrémens".[16] Moreover, the *abbé* strongly supported the idea of introducing children to physics by adequately adapting the teaching approach, because

> Un Enfant qui aura vû par forme de récréation les premiers principes démontrés d'une maniere capable d'intéresser sa curiosité, se portera de lui-même aux applications pour le peu qu'il soit aidé, & quand il sera tems de l'appliquer plus sérieusement à l'étude de la Physique, son esprit disposé de loin s'y livrera avec moins de peine, & plus de succés.[17]

[16]Nollet (1738, pp. xxviii–xxix) ("It is no more believed that knowing something that could be definitely ignored should be a reason to flush; it is well known that an enlightened spirit is not incompatible with other attractions").

[17]Nollet (1738, pp. xxxii–xxxiii) ("Having seen as a form of fun the first principles demonstrated in a way that aroused his curiosity, a child will get by himself closer to the applications with just a little help, and when it will be time to engage him more seriously in the study of physics, his spirit, having been predisposed for a long time, will devote itself to the task with less fatigue and more success").

It is within this context that Laura Bassi started practising experimental physics, acquiring such a reputation that Nollet was delighted to meet her in 1749, when he travelled through Italy, and began corresponding with her. The French scientist dedicated to her one of his *Lettres sur l'électricité,* that were published in three volumes from 1753 to 1767. In these letters, Nollet presented his most recent experiments and defended his theories on electricity.[18] The letter to "Madame Laura Bassi de l'Académie de l'Institut de Bologne" focused on some curious applications of electricity, that Nollet suggested to include in the lectures of experimental physics that Bassi held at home. According to Nollet, these demonstrations could be further improved but, as he underlined in his letter to the erudite lady, "en vous faisant part de ces premières tentatives, mon dessein est d'exciter votre émulation, persuadé que personne n'est plus capable que vous d'augmenter & de perfectionner ce que je n'ai fait qu'ébaucher".[19] He added that: "un motif encore plus puissant m'a déterminé à vous offrir cette partie de mon travail, c'est le désir de rendre mon hommage à une Dame qui a mérité celui de toutes les personnes appliquées aux sciences, & qui fait tant d'honneur à la Physique, par ses talents & par ses vertus".[20]

It is worth analyzing in more detail the lectures proposed by Nollet himself. He thoroughly describes these in his *Leçons de Physique Expérimentale*, a treatise in six volumes, full of accurate illustrations of experiments and instruments, published from 1743 to 1764. The text is remarkable for its clarity and its strict structure. For every topic, after an introduction giving the current knowledge on the question, Nollet moves on to the experiments, presenting systematically, for each experiment, in distinct subsections: the "preparation"—he describes the instruments used and the way the experiment is carried out— the "effects", the "explanation" and the potential "applications", *i.e.* the connections with everyday life or natural phenomena (Fig. 3.6). Nollet's lectures included the same classic topics as his predecessors, but with a particular attention to electricity, a field he was one of the major exponents of at the time (Fig. 3.7). A large part of one of the six volumes of the *Leçons de Physique Expérimentale* is actually dedicated to the study of electric science and Nollet published several articles and treatises focusing exclusively on electricity.[21] The French scientist also paid great attention to the details of his way of teaching. He wondered for instance whether it was worth preparing some "cahiers"—"lecture notes" in modern terms –, in order to read or recite them. However, as he himself objected, "Qui est-ce qui ne sçait pas que les meilleures choses données de cette manière endorment à la fin les Auditeurs, & ne se concilient que rarement l'attention

[18]Nollet (1753–1767). The "vingt-deuxième lettre Qui contient quelques applications curieuses de l'Electricité, à Madame Laura Bassi" is in the third volume, published in 1767, pp. 274–295.

[19]Nollet (1753–1767, vol. 3, p. 295) ("by presenting to you these first attempts, my aim is to excite your emulation, as I am sure that no one is more skilful than you to develop and improve what I have only sketched").

[20]Nollet (1753–1767, vol. 3, p. 295) ("a more powerful reason led me to offer to you this part of my work, it is the desire to pay my homage to a Lady who has deserved the homage of all those persons interested in sciences and has honoured Physics thanks to her skills and her virtues").

[21]Nollet (1745, 1747, 1749).

Fig. 3.6 Illustration of different models of pumps from Nollet's *Leçons de physique expérimentale* (Guérin, Paris 1743–1764)

qu'elles peuvent mériter?".[22] He concluded that it was better "de se former une habitude d'opérer en parlant, & même d'employer moins les paroles que l'exposition des faits [...] de façon que chacun, quand il voudroit faire des objections, & demander des éclaircissemens, n'eût point à craindre d'interrompre un discours étudié".[23]

Nollet's brilliant lectures involved the usage of about 350 instruments, a collection that Nollet set up despite many difficulties. As he explained, up to that moment, in France, instruments had hardly ever been used, so that there were no craftsmen able to design and make all the devices he needed. Buying the instruments from abroad

[22] Nollet (1738, p. xxii) ("Who does not know that the best things, when presented in this way, make the audience fall asleep and hardly ever catch the attention they deserve?").

[23] Nollet (1738, p. xxiii) ("to get into the habit of working and speaking at the same time, and using less words than the presentation of facts [...] so that nobody should be afraid of interrupting a prepared speech, if one wants to make objections and ask for clarifications").

Fig. 3.7 A lecture on electricity given by Nollet: the abbé, who stands on the left, shows the effects obtained by electrifying the suspended woman, insulated from the ground (Jean-Antoine Nollet, *Essai sur l'électricité des corps*, Guérin, Paris 1746)

would have been extremely expensive with respect to Nollet's means at the beginning of his career. So, Nollet wrote,

> J'ai pris moi-même la lime & le ciseau, j'ai formés & conduits des ouvriers pour m'aider; j'ai intéressé la curiosité de plusieurs Seigneurs qui ont placés de mes ouvrages dans leurs Cabinets; j'ai levé une espece de contribution volontaire; en un mot, je ne le dissimule pas, j'ai fait deux ou trois instruments d'une même espece afin qu'il pût m'en rester un.[24]

He later wrote an entire treatise, *L'art des experiences*, focusing on the description of instruments and instrument-making techniques (Fig. 3.8).

[24]Nollet (1738, pp. xviii–xix) ("I myself took a file and a chisel, I trained and guided workers so they could help me; I aroused the curiosity of several Lords who put my works in their own Cabinets; I obtained a sort of voluntary donation; in other words, I do not deny that I made two or three instruments of the same kind so that I could keep one for me").

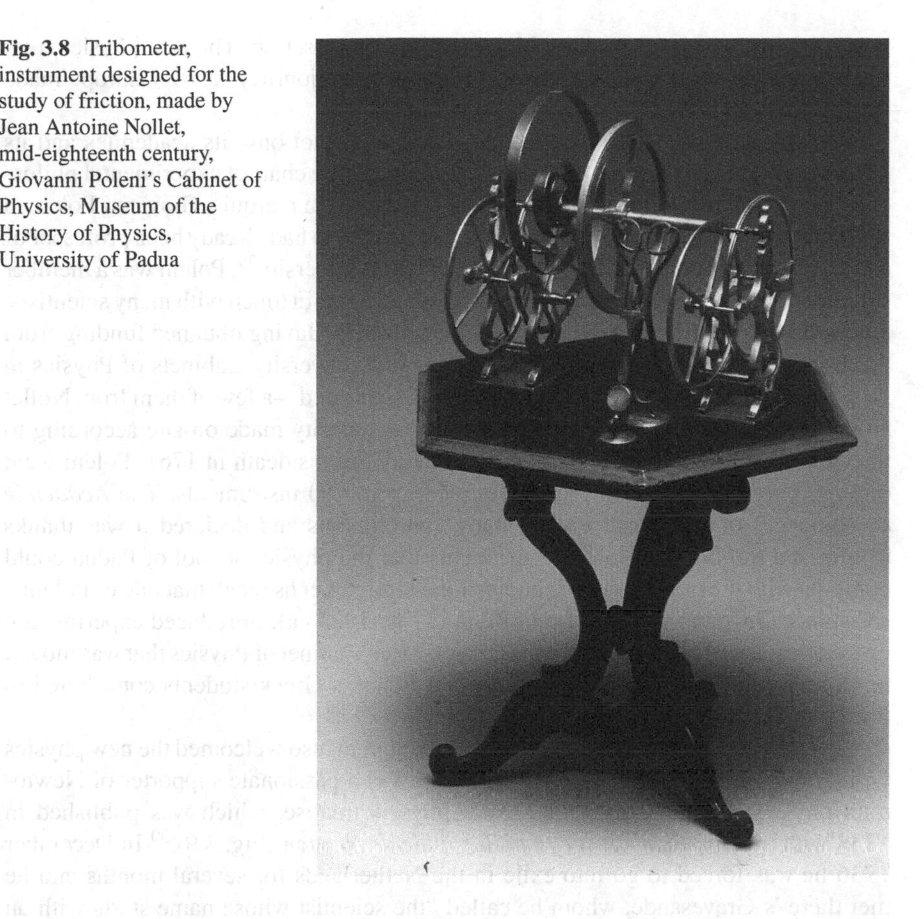

Fig. 3.8 Tribometer, instrument designed for the study of friction, made by Jean Antoine Nollet, mid-eighteenth century, Giovanni Poleni's Cabinet of Physics, Museum of the History of Physics, University of Padua

Nollet's lectures and treatises were extraordinarily successful and, along with his researches—above all the ones on electricity—, opened him the doors of a shining career. The scholar was invited to Turin in 1739 as teacher of the crown prince, he taught at the *Académie Royale de Bordeaux* in 1741, and three years later at the French court in Versailles, in the presence of the *Dauphin* and the Queen. He was assigned the chair of experimental physics, which had been just created, at the *Collège de Navarre* and became member of several academies, the *Académie Royale des Sciences*, of course, but also the Royal Society. He was also appointed *Maître de Physique* of the French king's children.

The lively activities carried out by Nollet and by the Dutch and English experimental physics pioneers brought forth striking results: in a few decades, the new physics lectures spread all over Europe, in universities as well as at Courts and in *Salons*. Chairs of experimental physics were created in many European universities and more and more Cabinets of Physics, homogeneous collections of scientific instruments for research and teaching purposes, were set up. The new physics treatises, the letters exchanged between scholars and the journeys across Europe made by the elite contributed to such a huge spread.[25]

Experimental physics lessons also reached Italy, not only its academies and its elite, but also its universities. In Padua, for instance, the chair of experimental philosophy was created in 1738. It was assigned to a Venetian marquis, Giovanni Poleni, a staunch supporter of the new experimental physics, who had already been professor of astronomy, philosophy and mathematics at Padua University[26]. Poleni was a member of several academies, both in Italy and abroad, and was in touch with many scientists, among whom Nollet and Pieter van Musschenbroek. Having obtained funding from the Republic of Venice, he set up one of the first university Cabinets of Physics in Italy.[27] Poleni bought some of the instruments he needed—a few of them from Nollet and Jan van Musschenbroek—, but he had the majority made on-site according to the experimental physics treatises of the time. Until his death in 1761, Poleni went on enriching his collection, which counted up to 400 instruments. The *Académie des Sciences* of Paris paid Poleni many compliments and declared it was thanks to him and to his collection of instruments that the physics school of Padua could compete with the most famous schools of the kind.[28] Let us recall that one of Poleni's assistants, Giovanni Antonio Dalla Bella (1730–1823 ca), introduced experimental physics teaching in Lisbon, where he created a rich Cabinet of Physics that was moved to Coimbra in 1772. Furthermore, some of Poleni's Greek students contributed to the introduction of experimental physics in Greece.[29]

Several philosophers of the Age of Enlightenment also welcomed the new physics with enthusiasm. Voltaire, for instance, who was a passionate supporter of Newtonian physics, started writing in 1735 a physics treatise, which was published in 1738 with the title *Eléments de la philosophie de Newton* (Fig. 3.9).[30] In December 1736 he was forced to go into exile in the Netherlands for several months and he met there 's Gravesande, whom he called "the scientist whose name starts with an apostrophe", and the Musschenbroek brothers. Back to France, he decided to collect physics instruments for experiments and demonstrations. He bought most of them from Nollet, spending a great amount of money, so that he wrote in October 1738: "L'abbé Nolet me ruine",[31] specifying in another occasion that "nous sommes dans

[25] See Bennett and Talas (2013).

[26] See Salandin and Talas (2000), Del Negro (2013) and Talas (2013).

[27] This is one of the first cases in Europe of a Cabinet of Physics which was totally publicly funded. As a matter of fact, the English lecturers worked, as we already mentioned, in collaboration with instrument makers, and 's Gravesande and Musschenbroek paid with their personal wealth the instruments they needed.

[28] Fouchy (1763).

[29] Talas (2004).

[30] Voltaire (1738).

[31] Letter from Voltaire to Thieriot, 27 October 1738, in Voltaire (1837, p. 290) ("the *abbé* Nollet is ruining me").

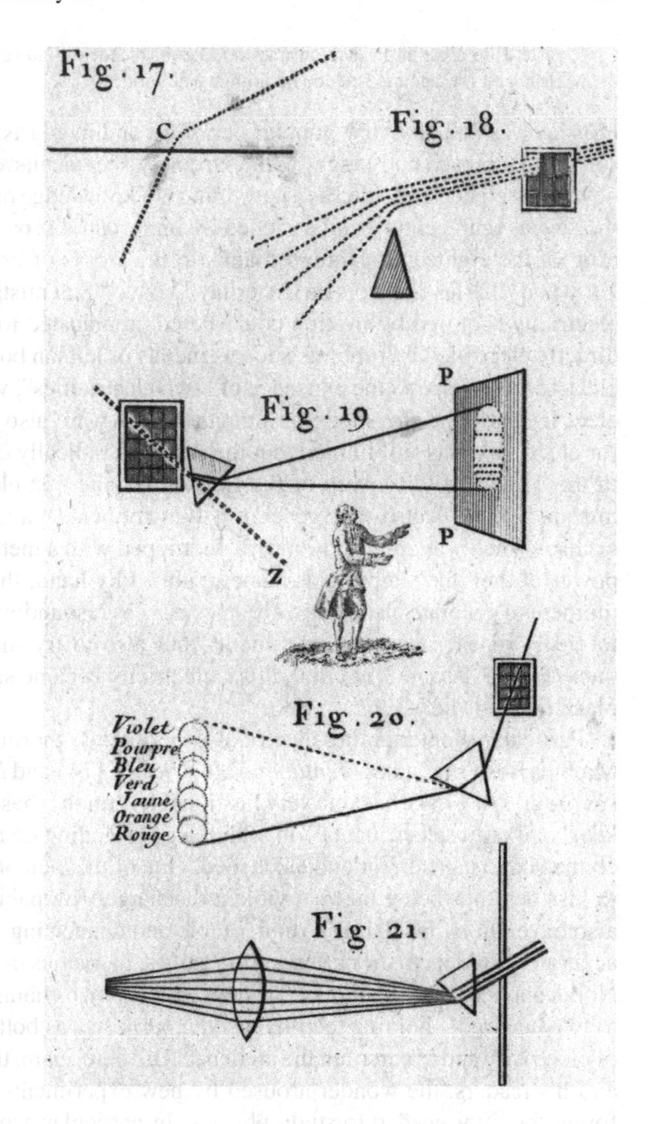

Fig. 3.9 Experiments on the decomposition of white light passing through a glass prism (Voltaire, *Eléments de philosophie de Newton,* in *Oeuvres complètes de Voltaire* , 70 vol., Kehl, Paris 1784–1789, 31, 1784)

un siècle, où on ne peut être savant sans argent".[32] The philosopher approached experimental physics very seriously and was afraid that, in the *Salons* of those years, physics would be just a passing trend:

> Les vers ne sont plus guère à la mode à Paris. Tout le monde commence à faire le géomètre et le physicien. On se mêle de raisonner. […] Ce n'est pas que je sois fâché que la philosophie [expérimentale] soit cultivée, mais je ne voudrais pas qu'elle devînt un tyran qui exclût tout

[32]Letter from Voltaire to Moussinot, June 1738, in Voltaire (1837, p. 269) ("we are living in a century, in which you cannot be a scientist without money").

> le reste. Elle n'est en France qu'une mode qui succède à d'autres et qui passera à son tour. Mais aucun art, aucune science ne doit être de mode.[33]

Physics was actually very popular in *Salons* and in courts. Electricity, in particular, surprised, amused and taught with extremely spectacular experiments. It was still a new born science, regarded as a new branch of knowledge only from the early years of the seventeenth century and enriched by important discoveries only from the beginning of the eighteenth century, thanks to the works of Francis Hauksbee, Stephen Gray and Charles François de Cisternay Dufay.[34] Scientists showed for instance that electricity acquired by friction could be communicated to bodies that could not be directly electrified by rubbing, such as metals or human bodies. They discovered the electric repulsion and the existence of "two electricities", vitreous and resinous. The electric spark and the shock on human beings were also highlighted. The interest for electricity was still limited, but the situation radically changed due to two events in the 1740s: the discovery of the Leyden jar, the first electric condenser, and the introduction of electrostatic generators—instruments made of a glass cylinder, globe or disc, which was spun and rubbed—equipped with a metallic "collector", far more powerful than the simple glass tubes, rubbed by hand, that had been used till that moment to generate electricity. The success was resounding. It was not only possible to easily repeat classical experiments, but also to try many new and spectacular ones (Figs. 3.10 and 3.11). In *Salons*, electricity became so popular that it took "the place of quadrille".[35]

Particularly entertaining electrical experiments, thoroughly described in Georg Mathias Bose's treatises (*Tentamina electrica* of 1744 and *L'électricité son origine et ses progrès* of 1754), became very fashionable. Among these was the famous "electric kiss" or "Venus electrificata", in which a girl standing on an insulated platform was connected to a generator and electrified. One of the men of the audience was invited to kiss her, thus being hit by a violent discharge with painful sufferings. One could also have fun by insulating a dinner table and connecting the electrostatic machine, accurately hidden, to the cutlery of the guests. In a series of well-known experiments, Nollet applied the discharge of Leyden jars to human chains of dozens of people—up to two hundred—holding their hands: the purpose was both to study the propagation of electricity and to entertain the audience. But here again, the *abbé* warned his public and his readers: the wonder aroused by new experiments should not make anyone forget the rigor needed to study physics. In particular, from the 1740s, the striking effects of electricity led scientists to try to apply it to the human body for medical purposes, and some experimenters claimed that electrical treatments gave prodigious

[33]Letter from Voltaire to Cideville, April 1735, in Voltaire (1837, p. 154) ("Verses are not fashionable any longer in Paris. Everyone starts being a geometer or a physicist. People are interested in thinking. [...] I do not regret that [experimental] philosophy is being studied, but I do not want it to become a tyrant who would exclude everything else. Here in France, it is just a trend that succeeds others and will vanish as well. But no art, no science should be fashionable").

[34]Heilbron (1979, 1999) and Peruzzi and Talas (2004).

[35]Haller (1745, p. 15).

Fig. 3.10 The electricity produced by the generator (on the right) is used to set fire to the alcohol contained in the spoon held by the lady on the left (William Watson, *Expériences et Observations, pour servir a l'Explication de la Nature et des Propriétés de l'Electricité*, Jorry, Paris 1748)

results. Nollet objected that experiments always needed to be carefully tested and repeated, and he disagreed with those who rushed into practical applications:

> Dès qu'il paroit quelque nouveauté en Physique, la curiosité s'en empare d'abord et s'en amuse, mais bien tot elle est satisfaite; elle fait place à l'intéret, & l'on exige que ce qu'on a admiré soit utile. L'impatience de certaines gens, à cet égard, va jusqu'à leur donner de l'humeur, & à leur faire regarder avec une sorte de mépris tout ce dont on ne voit pas d'abord une application à faire.[36]

It is worth underlining that the instruments which became classical elements of eighteenth century Cabinets of Physics had different characteristics. Only some of them enabled new experiments and innovative discoveries. This is the case, for instance, of electrostatic generators or air pumps, which allowed experiments in rarefied or compressed air. As for most instruments, these were typical teaching devices, intended to demonstrate and repeat phenomena already well-known at the time in the fields of mechanics, hydrostatics or optics. Let us recall that in those very years, the reproducibility of experiments was asserting itself as one of the main parameters of validation of the new physics.

Object of astonishment and wonder, physics reached in the eighteenth century one of the highest levels of popularity in its history, and such a popularity undoubtedly played a crucial role as for the great developments that were to take place

[36]Nollet (1746, pp. 18–19) ("As soon as some news comes out in physics, curiosity gets hold of it and enjoys it, but it soon gets satisfied; it gives place to self-interest and people demand that what has been admired has to be useful. The impatience of some people, in this sense, even makes them angry and they regard contemptuously whatever does not show a possible immediate application").

Fig. 3.11 The Leiden experiment. The water contained in the bottle is electrified thanks to the globe friction generator on the right. If the man who holds the bottle touches the metallic collector with his other hand, he is struck by a violent discharge (Jean- Antoine Nollet, *Essai sur l'électricité des corps*, Guérin, Paris 1746)

in physics in the nineteenth century. As a matter of fact, Enlightenment not only brought forth a notable increase in the number of "professional" physicists, but the passion and enthusiasm that marked physics in those years certainly contributed to the emerging of an outstanding generation of scientists at the beginning of the nineteenth century, with figures like Biot, Foucault, Faraday, Ampère or Fresnel. Moreover, the careful attention paid in the eighteenth century to the teaching and popularization of science definitely contributed to sensitize society to science so that France, for instance, during the revolution period, while it was isolated and at war with the rest of Europe, decided to invest huge resources in science and technology by sponsoring scientific magazines, financing patents and strongly encouraging the collaboration between scientists and scientific instrument makers. From that moment onwards, French science entered into an astounding fruitful period and the French precision industry dominated the European market during a large part of the nineteenth century. The Enlightenment thus teaches us that a serious and diffused sensitization to science gives great results, a lesson which is still relevant today, while the lack of scientific education—Nollet underlined it with emphasis—explains why there are still so many people who are "livrées à toutes les erreurs populaires, préoccupées des craintes les plus ridicules, susceptibles de tout le faux merveilleux & de toutes les charlataneries dont on voudra se servir pour les tromper".[37]

References

Jim Bennett, Sofia Talas (eds), *Cabinets of Experimental Philosophy in Eighteenth-Century Europe*, Brill, Leiden-Boston 2013

Peter De Clercq, *At the sign of the Oriental Lamp. The Musschenbroek workshop in Leiden, 1660–1750*, Erasmus Publishing, Rotterdam 1997

Peter De Clercq, *The Leiden Cabinet of Physics*, Museum Boerhaave Communications, Leiden 1997

Piero Del Negro (ed.), *Giovanni Poleni tra Venezia e Padova*, Venice 2013

John Theophilus Desaguliers, *A course of experimental philosophy*, 2 vol., Senex, London 1734–1744

Jean-Paul Grandjean de Fouchy, *Eloge de M. le Marquis Poléni*, «Histoire de l'Académie Royale des Sciences», 1763, pp. 151–163

Willem's Gravesande, *Physices elementa mathematica, experimentis confirmata, sive introductio ad philosophiam Newtonianam*, Vander Aa, Leiden 1720–21; 2nd ed. 1725; 3rd ed., Langerak e Verbeek, Leiden 1742

Willem 's Gravesande, *Mathematical Elements of Natural Philosophy, Confirm'd by Experiments: or, an Introduction to Sir Isaac Newtons' Philosophy*, Senex & Taylor, London 1720–21

Albrecht von Haller, *Histoire des nouvelles découvertes faites, depuis quelques années en Allemagne, sur l'électricité*, «Bibliothèque raisonnée des ouvrages des savans de l'Europe», XXXIV, 1745, pp. 3–20.

John L. Heilbron, *Nollet, Jean Antoine*, in *Dictionary of Scientific Biography*, 16 vol., Charles Scribner's Sons, New York 1970-1980, vol. 10

John L. Heilbron, *Electricity in the 17th and 18th centuries*, University of California, Berkeley 1979; Dover publications, New York 1999

Alan. Q. Morton, Jane A. Wess, *Public and Private Science – The King George III Collection*, Oxford University Press, Oxford, and Science Museum, London 1993

Pieter van Musschenbroek, *Elementa physicae*, Luchtmans, Leiden 1734

Pieter van Musschenbroek, *Introductio ad philosophiam naturalem*, 2 vol., Luchtmans, Leiden 1762

Jean Antoine Nollet, *Programme ou idée générale d'un cours de physique expérimentale*, Le Mercier, Paris 1738

Jean Antoine Nollet, *Leçons de physique expérimentale*, 6 vol., Guérin, Paris 1743–1764

Jean Antoine Nollet, *Conjectures sur les causes de l'électricité des corps*, «Mémoires de l'Académie Royale des Sciences de Paris», 1745, pp. 107–151

---

[37]Nollet (1738, p. xxxvi) ("are at the mercy of all popular mistakes, worried by the most ridiculous fears, subject to any wonderful falsehood and to all the quackeries that may be used to deceive them").

Jean Antoine Nollet, *Observations sur quelques nouveaux phénomènes d'électricité*, «Mémoires de l'Académie Royale des Sciences de Paris», 1746, pp. 1–23
Jean Antoine Nollet, *Eclaircissemens sur plusieurs faits concernant l'électricité*, «Mémoires de l'Académie Royale des Sciences de Paris», 1747, pp. 102–131
Jean Antoine Nollet, *Recherches sur les causes particulières des phénomènes électriques*, Guérin, Paris 1749
Jean Antoine Nollet, *Lettres sur l'électricité*, 3 vol., Guerin & Delatour, Paris 1753–1767
Jean Antoine Nollet, *L'art des expériences*, 3 vol., Durand, Paris 1770
Giulio Peruzzi, Sofia Talas, *Bagliori nel vuoto – Dall'uovo elettrico ai raggi X: un percorso tra elettricità e pneumatica dal Seicento a oggi*, Edizioni Canova, Treviso 2004
Pierre Polinière, *Expériences de Physique*, Jean de Laulne, Paris 1709
Gian Antonio Salandin, Sofia Talas, *Strumenti scientifici, dalla raccolta di Vallisneri al Teatro di filosofia sperimentale di Poleni*, in *La curiosità e l'ingegno,* Università di Padova, Padova 2000, pp. 222–243
Gian Antonio Salandin, Sofia Talas, *Giovanni Poleni*, in *La curiosità e l'ingegno*, Università di Padova, Padova 2000, pp. 84–91
Sofia Talas, *The creation and the role of Giovanni Poleni's Teatro di Filosofia Sperimentale*, in *Bisanzio, Venezia, Nuovo Ellenismo*, ToDe Publications, Athens 2004, pp. 283–293
Sofia Talas, *New Light on the Cabinet of Physics of Padua*, in Jim Bennett and Sofia Talas (eds), *Cabinets of Experimental Philosophy in Eighteenth-Century Europe*, Brill, Leiden-Boston 2013, pp. 49–67
Sofia Talas, *Il gabinetto di filosofia sperimentale di Poleni*, in Piero Del Negro (ed.), *Giovanni Poleni tra Venezia e Padova*, Venice 2013, pp. 247–275
Voltaire, *Éléments de la philosophie de Newton*, Ledet & cie, Amsterdam 1738
Votaire, *Œuvres complètes*, vol. 11, Furne, Paris 1837

Chapter 4
Always Among Men: Laura Bassi at the Bologna Academy of Sciences (1732–78)

Paula Findlen

On 20 March 1732 Laura Bassi, daughter of a Bolognese lawyer and "a young girl, nineteen years old", became the first female member of the Accademia dell'Istituto delle Scienze di Bologna. The sixteen academicians present for this meeting unanimously agreed to her admission "to the ranks of the honorary academicians" after hearing

> the presentation that Signor Eustachio Manfredi, Signor Beccari, Father Abundio Collina, and others gave regarding the infinite and incredible erudition demonstrated by this young girl, beyond her sex and age, supported by the many conclusions that she sustained many times about all of philosophy, with such liveliness, quickness, nobility of speech, and profound learning that you would not be able to believe it if you had not heard her.

Two of the academy's most important members, the Istituto professor of physics Jacopo Bartolomeo Beccari and the academy secretary Francesco Maria Zanotti, were deputized to inform Bassi so that she would know of the "high esteem that the academy had of her intelligence".[1]

The Istituto's decision to make Bassi an academician played an important role in her emergence as one of the most famous and admired women of science in the eighteenth century. One month later, her spectacularly well-publicized defense of forty-nine philosophical theses on 17 April 1732 qualified her for a *laurea* in philosophy

[1] Archivio dell'Accademia dell'Istituto delle Scienze, Bologna (hereafter AASB), *Registro degli Atti dal 1723 al 1803*, n. 5 (20 March 1732). See also informal reports of this event such as Biblioteca Comunale dell'Archiginnasio, Bologna (hereafter BCAB), B. 382, letter 32 (Giampietro Zanotti to Father Giampietro Riva, Bologna, 9 April 1732). Bassi has been the subject of a considerable literature in recent years. See Paula Findlen (1993), Gabriella Berti Logan (1994), Beate Ceranski (1996), Marta Cavazza (2006, pp. 61–85; 2020). Many other publications by Cavazza, whose work has been fundamental to this subject, will be cited throughout the notes.

P. Findlen (✉)
Department of History, Stanford University, Stanford, CA, USA
e-mail: pfindlen@stanford.edu

L. Cifarelli and R. Simili (eds.), *Laura Bassi–The World's First Woman Professor in Natural Philosophy*, Springer Biographies,
https://doi.org/10.1007/978-3-030-53962-7_4

which was awarded in May. A well-paid professorship in universal philosophy at the University of Bologna followed at the end of October. In the publicity surrounding these later events, it is easy to forget that the Istituto took the lead in creating a special place for her in the scientific culture of Bologna. For almost forty-six years, until her death in February 1778, Bassi would take advantage of every opportunity that academy membership offered her to cultivate her reputation as an accomplished and knowledgeable physicist.

4.1 Laura Bassi, Accademica Onoraria (1732–45)

The admission of Laura Bassi was one of the events that marked the Istituto's transition from the first to the second phase of its development. In 1732 Luigi Ferdinando Marsigli, the ambitious, prickly, and demanding founder of the Istituto, had been dead less than two years. His perpetual dissatisfaction with the Bolognese Senate's execution of the terms of his bequest, the facilities in Palazzo Poggi, and the work of the Istituto professors and academicians had almost destroyed the temple of learning he inaugurated in 1711. Were it not for the timely and diplomatic intervention of Prospero Lambertini, then bishop of Ancona, the Istituto might have ceased to exist altogether around 1726.[2] But Marsigli was no longer around to meddle in the Istituto's affairs and threaten to withdraw his support. Instead, Lambertini's return to his native city in 1731, as Archbishop of Bologna after many years in Rome, gave the Istituto a powerful and supportive patron who aspired to realize the program that Marsigli could never quite execute. Under the editorial supervision of the Istituto secretary Francesco Maria Zanotti, the long delayed first volume of the Istituto *Commentarii* finally appeared in 1731.[3] Things were indeed looking up.

How did Bassi fit into these plans? From the start, key Istituto professors and academicians were involved in her education, though they did not all agree on how she should be educated. Her philosophy tutor and family physician Gaetano Tacconi had been a member of the academy since 1717.[4] After persuading Giuseppe Bassi to invite learned scholars into his home to engage in philosophical debate with his daughter and observe her talents, Bassi's learning became widely known and admired. While we have no record of who attended the many evenings of philosophical conversation in Casa Bassi, all the evidence suggests that other members of the Istituto academy were among the most important participants in these debates and began to direct her attention to subjects Tacconi had not covered in his private lessons.[5] Thus, prior to her April defense, Bassi's knowledge had been tested repeatedly by key figures such as

[2]On the history of the Istituto delle Scienze, see Marta Cavazza (1990), Walter Tega (1986, 1987), Annarita Angelini (1993), John Stoye (1994).

[3]Mauro De Zan (1990).

[4]Richard Leonard Rosen (1971, p. 191).

[5]According to Giampietro Zanotti, Bassi presented her philosophical theses "privatamente in Casa sua cinquanta volte," which suggests how frequent the opportunities were for key Institute members

Zanotti, Beccari, Eustachio and Gabriele Manfredi, Domenico Gusmano Galeazzi, Matteo Bazzani, and probably the youngest and newest members, Francesco Algarotti and Eustachio Zanotti. They would be among those admirers who urged Cardinal Lambertini to visit Casa Bassi to witness the erudition of *la donzella Laura.*

Lambertini's decision to have the university to make Bassi's scientific learning a public fact in the city was largely executed by Istituto members, many of whom were university professors while others belonged to the urban aristocracy and clerical elite of the city. As Rector of the College of Arts, Bazzani presided over her degree ceremony on May 12, where he gave an oration in praise of Bassi and specifically complemented Tacconi for his education of this learned woman.[6] Beccari and Gabriele Manfredi examined her on matters of natural philosophy and physics. In a lost oration that we only know by its title, *De philosophicorum studiorum utilitate in mulieribus desiderabilium Oratio*, Tacconi seems to have presented Bassi as the ultimate demonstration not only of a woman's right to education but of the Cartesian philosophy of mind and body.[7] This formulation of Bassi's learning sparked a bitter disagreement between Tacconi and other academicians about the future direction of her studies that would force Bassi to make a choice. Bazzani's diplomatic praise of Tacconi's role as *maestro* masked real intellectual differences within the ranks of the academicians in which she was about to be embroiled.

The more mathematically inclined and experimental academicians, all self-professed Newtonians, felt that Bassi had the ability to pursue natural philosophy beyond general qualitative questions about the nature of things that belonged to the long tradition of learning stretching from Aristotle to Descartes. Her brief exposition of Newton's optics in her physics theses, which became the centerpiece of Algarotti's poetic exultation of Bassi as "the shadow of the Great British", hinted of what was to come.[8] The Newtonian circle within the academy openly discussed the deficiencies of Tacconi's course in philosophy and derided the largely scholastic content of the physical theses he had proposed for her public defense. His subsequent decision to encourage Bassi to discuss questions of ethics for her first philosophical debate after her degree spurred them into action. They encouraged Lambertini to intervene in favor of a more scientific subject of longstanding interest to Bologna's scientists: the nature and composition of water. Bassi presented twelve theses on this subject on 27 June 1732 and drafted other unpublished theses that reveal a growing engaging with Newtonian science.[9] By shaping the content of Bassi's public presentations of her learning, once her degree had been conferred and she was well on her way to becoming a university professor, the Newtonian members of the Istituto established her status as a modern natural philosopher.

---

to get to know her well. BCAB, B.382, letter 34 (Giampietro Zanotti to Father Giampietro Riva, Bologna, 22 June 1732).

[6]Matteo Bazzani, *Oratio … ad Egregiam Virginem D. Lauram Mariam Cattarinam Bassi* (1732) in Giovanni Fantuzzi (1778, pp. 31–32).

[7]Michele Medici (1864).

[8]Francesco Algarotti (1732) and Massimo Mazzotti (2004).

[9]Gabriella Berti Logan (1999, pp. 499–501).

Beccari, Manfredi, and Zanotti—and from a distance the papal physician and ardent Newtonian Antonio Leprotti—felt that Bassi had a mind made for modern physics. They encouraged her to pursue this subject, acquiring the knowledge necessary through a broader reading of recent works of modern natural philosophy, and to perfect the mathematical skills necessary for such an undertaking. Even as she accepted a special lectureship in "universal philosophy", her contact with energetic and supportive members of the Istituto academy led Bassi to abandon the philosophical path outlined in her course of studies with Tacconi. She would continue to use this knowledge as the elementary foundation of her introductory philosophy lectures but her own interests began to converge with the intellectual agenda of the most ambitious and widely read members of the Istituto. Bassi made plans to study algebra with Gabriele Manfredi who would tutor her in this subject for three or four years until she became proficient in differential calculus—a rare commodity even among physics professors in mid-eighteenth century Italy.[10] She seems to have also considered Beccari an intellectual mentor, describing him as her "most learned master" at the time of his death.[11] The Cartesian anatomist Tacconi may have introduced her to members of the Istituto but it quickly became clear that the Newtonian avant-garde of the academy adopted her as their intellectual project. She was a far bolder and clearer demonstration of the success of a Newtonian worldview than the cautiously worded account of Algarotti's and Zanotti's successful replication of Newton's prism experiments in the *Commentarii*, not a qualitative *Newtonianism for Ladies* as the famous title of Algarotti's 1737 book suggested but a quantitative Newtonianism produced by a young and highly learned woman.[12]

This is how Laura Bassi began her relationship with the Istituto delle Scienze. Within a matter of months, the Istituto added a second female member at a distance when they admitted the Neapolitan noblewoman Faustina Pignatelli on 20 November 1732, "having the most certain testimonials of this lady's great and marvelous worth in mathematics, and especially algebra". They did this with the explicit understanding "to not accept any other woman into the academy".[13] There was surely some concern that they were now setting a precedent that might lead to the admission of every scientifically inclined woman in Italy, including local noblewomen such as Laura Bentivoglio Davia and Elisabetta Ercolani Ratta who had both studied natural philosophy and mathematics with Zanotti, or possibly the Manfredi sisters Maddelena and Teresa who were well-known for their astronomical calculations.[14] Pignatelli already

[10]Archivio di Stato, Bologna (hereafter ASB), *Assunteria di Studio. Requisiti dei lettori*, vol. 16, n. 27 (Gabriele Manfredi, 19 December 1737). "Per quattro'anni che spirano ha avuto la sorte di essere Compagno di Studio nelle Materia Algebriache della dottissima Sig.a Laura Bassi". Cf. *Requisiti dei lettori* (Laura Bassi, 1739) in which she describes their lessons as lasting three years.

[11]Biblioteca Apostolica Vaticana, *Aut. Patetta*, cart. 45 (Bassi to Giambattista Beccaria, n.d.).

[12]On this subject, see Mauro De Zan (1990).

[13]AASB, *Registro degli Atti dal 1723 al 1803*, n. 5 (20 November 1732).

[14]Ilario Magnan Campanacci (1988), Paula Findlen (2005). My forthcoming study of the making of Algarotti's *Neutonianismo per le dame* (1737) will discuss Elisabetta Ercolani Ratta in some detail.

demonstrated the kind of mathematical competency to which Bassi aspired. She earned her membership because of her elegant solution to a mathematical problem posed by Eustachio Manfredi presented by Pietro di Martino, the younger brother of her mathematics tutor Nicola who was then a student in Bologna. She further confirmed the justness of this decision when her comments on the problem of *vis viva* were published anonymously in the Leipzig *Acta eruditorum*, one of the most important scholarly journals of the day that also included a notice of Laura Bassi's degree and professorship.[15] Pignatelli would become one of Zanotti's intimate correspondents and a lifelong supporter of the Istituto, even though she never traveled to Bologna. Despite Bassi's friendship with di Martino, whom she met during his stay in Bologna, there is no record of any contact between these two women.[16] Only in retrospect did the Istituto perceived Bassi's admission as the beginnings of their efforts to create a community of women celebrated by the Istituto delle Scienze.[17]

On an institutional level, Bassi's academy membership remained purely ceremonial, adding luster to both the woman and the Istituto academy without any expectation that she would participate in the more mundane activities of the academy. The category of academician in which she had been placed (*Onorarii*) initially had been reserved for nobles and foreigners, and subsequently was restricted only to foreign members in 1722. Ten years later, it became a category for women as well as foreigners.[18] Yet even limited membership had its benefits. Bassi's role as an academician gave her a recognizable place in the European-wide republic of letters as a member of a community defined not only by local interests but also by perceptions of the value and utility of scientific knowledge throughout the learned world.[19] Foreign scholars who came to Bologna to see the Istituto often met Bassi while those who became corresponding members of the academy were aware of her activities and, in some instances, engaged her in correspondence as a sign of respect and admiration. Even academicians who did not write directly to Bassi were aware of her place in this world. In the archives of the Paris Academy of Sciences, there is a draft letter written by the Paris Academy secretary Jean-Paul Grandjean de Fouchy to his counterpart in Bologna, Zanotti. Thanking Zanotti for a copy of the most recent *Commentarii*, he also acknowledged the additional copies that Bassi included with Zanotti's letter to be distributed to other French colleagues, noting on the back of the

[15]On Bassi's degree, see the «Nova acta eruditorum», n. 7, July 1732, pp. 341–343, Pignaetelli's solutions to four mathematical problems, including her rethinking of Leibniz's account of *vis viva*, were published anonymously though the identity of the *Anonyma Neapolitana* was an open secret: *Problemata mathematica Neapoli ad Collectores Actorum Eruditorum transmissa*, «Nova acta eruditorum», n. 1, 1734, pp. 28–34.

[16]AASB, *Missive Zanotti*, 156 (Francesco Maria Zanotti to Pietro de Martino, Bologna, 4 January 1741).

[17]Marta Cavazza (2000).

[18]Richard L. Rosen (1971, pp. 67–69). As of 1722, *Onorari* was revised to describe only foreign members. Pignatelli was therefore doubly "honored", as a foreign woman. In 1735, a subsequent simplification of the early system divided academicians into two classes: *Ordinarii* and "tutti gl'Altri." AASB, *Registro degli Atti dal 1723 al 1803*, n. 6 (18 April 1735).

[19]James E. McClellan III (1985).

letter that this was correspondence regarding "Lady Laura Bassi".[20] In this fashion, the Paris academicians acknowledged the existence of *la filosofessa di Bologna*.

Bassi attended no academy meetings, public or private, before November 1745. She presented none of her research to the academy until April 1746. For over a decade she had no actual presence in the institution that had launched her career. Her marriage to fellow academician Giuseppe Veratti in February 1738 did nothing to alter this fact. Veratti, a pupil of Beccari, first presented his research before the Istituto in December 1733 and became an academy member in April 1734, *Ordinario* and Vice-President in 1742, and President in 1743.[21] Bassi had indeed married one of the active and ambitious younger academicians although Veratti would not hold an Istituto professorship until her death in 1778.

Despite the glorious celebration of Laura Bassi in 1732, throughout the 1730s there were ongoing concerns about whether the academy would continue to exist. "We discussed the wretched state of the academy, and the lack of subjects on which some dissertation is recited at the predetermined times", recorded secretary Zanotti in July 1733.[22] Sporadic discussions about the need for a more dynamic and engaged membership as well as the importance of sponsoring public events that would bring greater recognition to the Istituto occurred throughout this decade. Yet even the Assunteria dell'Istituto's request in November 1738 to have "more frequent semi-public academies at which we hear those whom we usually never hear" did not prompt the academicians to think of inviting Bassi to give a public lecture, or even to participate in more ordinary events such as the collective reading of the *Acta eruditorum* and *Mémoires* of the Paris Academy of Sciences that periodically occupied entire meetings.[23] While no one ever explicitly stated that she was not invited, it was tacitly understood that her presence was not required.

How did Laura Bassi feel about the honorific nature of her academy membership? For much of her teaching career, she actively petitioned the university to expand the scope of her teaching duties which had been defined, *ratione sexus*, as largely ceremonial. Her numerous petitions to the Senate and her efforts to create a private school make it clear that she aspired to teach. After a failed attempt to transform her lessons with Manfredi into a mathematics course, Bassi finally succeeded in creating a regular teaching venue by introducing a course in experimental physics in her home in 1749. The fact that the Istituto also had a chair in physics—held by Beccari until 1734 and then Galeazzi until 1770, both supporters of Bassi—was not unrelated to this choice. The Istituto professor lectured for two hours weekly while Bassi offered daily lessons. Both had machines with which to demonstrate experimental physics,

[20] Archive de l'Académie des Sciences, Paris, *Dossier Zanotti F. M.* (Jean-Paul Grandjean de Fouchy to Francesco Maria Zanotti, February 1753).

[21] AASB, *Registro degli Atti dal 1723 al 1803*, n. 5 (20 November 1732), AASB, *Registro degli Atti dal 1723 al 1803*, n. 5 (17 December 1732, 2 and 8 April 1734); n. 8 (8 November 1742), Richard L. Rosen (1971, pp. 173–175). Veratti would hold the office of Vice President several times and again serve as President in 1756. On their marriage, see Beate Ceranski (1996, pp. 89–95) and *passim*; Paula Findlen (2003), Marta Cavazza (2009).

[22] AASB, *Registro degli Atti dal 1723 al 1803*, n. 5 (6 July 1733).

[23] AASB, *Registro degli Atti dal 1723 al 1803*, n. 8 (13 November 1738; 21 March 1743).

but Bassi's proficiency in mathematics made her far more capable of explaining Newtonian physics than the physician Galeazzi or his assistant Paolo Balbi. She also developed an experimental regimen that made her highly proficient in the use of machines for teaching as well as research. Finally, it is clear that Galeazzi found his duties to experimental physics a burden that took him away from his obligations as a practicing physician and did not bring in enough income to justify the time he spent demonstrating experiments to students, nobles, and distinguished foreigners.[24] While Bassi did not explicitly challenge Galeazzi's position as the Istituto professor of physics, she simply offered a better and more regular physics course than either the university or the Istituto could provide, and did so with great enthusiasm and dedication. In such decisions, we see the seeds of a project that would eventually qualify her to become the Istituto professor of experimental physics in 1776.[25]

These were Bassi's decisions about how to address the limits of her teaching opportunities. But they do not answer the question about how she felt about her exclusion from academy meetings. To some degree, this omission was in name only. Many of the key academicians were regular participants in philosophical conversations in Bassi's home, and debated her publicly in the annual Carnival anatomy image where she demonstrated a broad scientific knowledge that went well beyond mathematical physics to include anatomy and physiology. Yet I think we can discern a hint of her desire to alter the situation in her request for a raise in 1739. Among the reasons justifying her request, Bassi mentioned her decision to inaugurate a biweekly private academy: "Since last year she has introduced an academy or learned conference in her home in which, two nights a week, she does philosophical and geometric exercises, etc."[26] Thus, in the year after her marriage to Veratti and in a period in which the Istituto continued to have difficulties persuading its members to meet regularly, Bassi created an alternative philosophical academy.

Bassi's evening *conversazioni* would continue to be an essential feature of her intellectual presence in Bologna along with her daily physics instruction. They served as a venue in which local scholars mingled with nobles and foreigners who were interested in natural and experimental philosophy and eager to meet Bassi. Let us keep in mind that the first thing that Abbé Nollet, the famous French electrical experimenter, did upon arriving in Bologna in 1749 was to have Zanotti bring him to Bassi's home where he subsequently returned for another evening. Unlike his problematic encounter with Veratti, whose experiments with medical electricity Nollet found entirely unsatisfactory, Nollet's experience of Bassi reinforced his sense that she was his most important colleague in Bologna.[27] In short, the status of Bassi's *conversazioni* continued to be an important part of her reputation even after she was offered the opportunity to attend academy meetings.

---

[24]Biblioteca Comunale, Forlì, *Autografi Piancastelli* 549.189 (Galeazzi to Leprotti, 10 July 1745).

[25]Marta Cavazza (1995, pp. 715–753).

[26]Elio Melli (1960, p. 87).

[27]On Nollet's visit to Bologna, see Paola Bertucci (2007).

4.2 "The Business of Signora Laura" (1745)

In 1740, after the longest conclave in living memory, Prospero Lamberini became Pope Benedict XIV (1740–58). He did not forget his native city—indeed he remained archbishop until 1754—nor did he neglect the Istituto. With his *Motu proprio* on 22 June 1745, he formally inaugurated a new class of Istituto academicians known as the *Benedettini*. The next two decades were a golden era for the Istituto. After becoming pope, Lambertini finally had the resources to invest properly in the renovation of science in the Papal States. There was great anticipation in Bologna as to what he might accomplish on their behalf. Beccari wrote to the papal physician Leprotti that he especially hoped to see Benedict XIV's papacy "do some good for this dead academy of ours that could get going again with a little assistance and, once revived, would permanently establish philosophical erudition in this land to the honor of the state and all of Italy".[28] Galeazzi made sure the pope knew just how dated the physics cabinet was, rendering it useless for teaching.[29] During the next few years the pope filled the Istituto library with books, encouraged the anatomist Ercole Lelli to work on an ambitious program of wax anatomies, and made sure that the physics cabinet was furnished with the best Dutch and English instruments, expanding it to fill three rooms. In 1742 Galeazzi wrote that the Istituto was now on the verge of having everything it needed for "a complete and perfect course in experimental physics".[30] He warmly joked that every time he did an experiment, he would think of the Holy Father.

The results were apparent to everyone, including the pope. In November 1744 Benedict XIV proudly referred to "the magnificent state of the Institute, the well-preserved machines given to the room for experiments, and Lelli's superb work, all of which have relieved me from my melancholy".[31] He completed his restoration of the Istituto by assigning it the income of the recently suppressed Collegio Panolini, which produced a *rendita annua* of 3600 lire that gave the Istituto greater financial autonomy from the Senate. The pope designated 2400 lire for the maintenance of the Istituto, and reserved the remaining 1200 lire for the creation of 24 *Benedettini*. Each would receive an annual 50 lire stipend in return for regular participation in the academy meetings and public presentation of at least one annual dissertation on their research. Fourteen positions were reserved for the Istituto president, secretary, professors, and their assistants. The other ten would be elected by the first group from among the members at large. Beccari wrote that he was less enamored by the pope's desire to create "raises for the Institute Professors", since it took money away

[28] Biblioteca Lancisiana, Rome, *Fondo Leprotti*, ms. 282 LXXVII. 1. 15, f. 26 (Beccari to Leprotti, Bologna, 30 August 1741).

[29] Nadia Urbinati, *Physica*, in Walter Tega (1987, p. 123).

[30] Biblioteca Lancisiana, Rome, *Fondo Leprotti*, ms. 282 LXXVII. 1. 15, f. 184 (Galeazzi to Leprotti, Bologna, 23 June 1742). On the growth of the Istituto physics cabinet, see Marta Cavazza (1995, pp. 716–717).

[31] ASB, *Archivio Malvezzi Campeggi. Corrispondenza*, II/358 (= 2788 in Casagrande inventory), (Benedict XIV to Paolo Magnani, Rome, 14 November 1744).

from the Istituto but he nonetheless appreciated the pope's intention of stimulating a more active and visible research program.[32]

It was around this time that Voltaire wrote Bassi a letter indicating his desire to become a foreign member of the Istituto. "There is no Bassi in London and I would be happier to become a member of your academy of Bologna than those of the English, even if they produced a Newton. If your patronage allows me to obtain this title, which I greatly desire, my heartfelt gratitude will be equal to my admiration for you".[33] With Bassi's recommendation, the academy admitted the French philosophe on 14 January 1745. Her ongoing ability to represent the Istituto abroad and attract well-known foreigners to the city, served as a reminder that she continued to be one of the most successful decisions the Istituto had made.

In the spring of 1745, as the list of the first *Benedettini* was being drawn up in Bologna and sent to Rome for final approval, Bassi learned that she was not among the ten members at large. She very much wanted to be part of this new initiative, not only in acknowledgment of the merits of her past work but also because of the new opportunities it would provide. Writing confidentially to her close friend Flaminio Scarselli, secretary of the Bolognese ambassador in Rome and a fellow academician, Bassi urged him to encourage the pope to consider the possibility of an additional position. She understood the importance of presenting this initiative as a separate decision from the composition of the original list of candidates since she did not wish to be seen as competing for one of the twenty-four positions. "However, it would be at the pope's discretion to place me in this series as I was placed in the university, as an extraordinary member, that is, an additional one". Bassi also encouraged Scarselli to inform Benedict XIV of her active research program, highlighting the fact that she had "material ready to provide the academy with a few dissertations". To further bolster the case for a twenty-fifth position, she suggested that it would be better to acknowledge, from the start, her qualifications to participate in this new initiative to create a community of elite researchers. Bassi rightfully felt that if she were to present her research after the selection of the inaugural group of Benedictines was announced, she might incur the animosity of some members who would see her as angling for the first vacancy. She concluded this remarkable petition by expressing her desire to use her talents to fulfill "my obligations towards the academy".[34] It is here that we finally see Bassi offer a pointed comment about her status as *Accademica Onoraria*. She wanted to be recognized as an active contributor to the academy who had earned the right to a Benedictine stipend in addition to the other privileges and obligations that came with this new and prestigious form of academy membership.

The pope listened sympathetically to Scarselli's suggestion and agreed, based on his own knowledge of Bassi's work. Benedict XIV emended the *Motu proprio* to include a twenty-fifth position, using almost verbatim the very words Bassi had sent Scarselli to articulate the reasons for this decision. Where he got the extra 50 lire from

[32]Biblioteca Lancisiana, Rome, *Fondo Leprotti*, ms. 282 LXXVII. 1. 15, f. 32 (Beccari to Leprotti, Bologna, 14 July 1742).

[33]Ernesto Masi (1881, p. 166) (Voltaire to Bassi, Paris, 23 September 1744).

[34]Elio Melli (1960, pp. 103–104) (Bassi to Scarselli, 21 April 1745).

was never specified though presumably it meant that the Istituto would buy a few less books and machines. Bassi was now the sole *Benedettina.* In the official history of the Istituto published in 1751, Giuseppe Gaetano Bolletti noted Bassi's unique position as further proof of the pope's great admiration for her.[35] Bassi began to contemplate the best way to thank Benedict XIV, though she curiously did not produce what Scarselli recommended: a published scientific paper with a fine dedication.[36]

The newly formed *Benedettini* began to organize their program of research. On 25 August 1745, they met to review the *Motu proprio.* Bassi was absent from this meeting but her place in the new organization, "as a supernumerary", was duly acknowledged. "Occasionally Signora Laura Bassi is allowed the right to be able to speak on that day that she prefers".[37] She now had an official invitation to present her work. Later that fall Bassi attended her first meeting when the *Benedettini* met in Beccari's home on November 18 to continue discussions about the reforms the Istituto was undertaking. This was not a public event so her presence at this meeting behind closed doors was not simply ceremonial but seemed to indicate a new stage in her relationship to the Istituto. Or did it? Five days later, the *Benedettini* held another meeting, this time at the Istituto. According to Galeazzi, who was placed in the uncomfortable position of explaining to Rome what exactly had transpired, Bassi and Veratti had been invited to this meeting but did not attend due to bad weather. Those who were present at this meeting began to discuss procedures for electing new members. Re-reading the papal bull, they felt that it clearly stated that "Signora Laura Bassi should be excluded from voting in the election of the Benedictines".[38]

When Bassi and Veratti heard about this decision, they were furious. Their unexpected absence had permitted a discussion about Bassi that surely would not have occurred in their presence. They skipped another meeting to consider how to handle the situation. By late November letters from both parties in this disagreement arrived in Rome, leaving the Istituto's patrons to contemplate the unintended effects of their generosity. Let us begin with Bassi's careful account of these events to secretary Scarselli. She debated whether to say anything at all but, in the end, felt that she could not let her fellow academicians interpret the meaning of a papal decree without offering her own reading of its intent regarding her status:

> It seems to me that the legitimate meaning of this additional position is that I may not be replaced, and nothing more. Since the academy did me the honor of admitting me, it gave me a vote and other common privileges, and equally in the college when doctorates are awarded, thus I do not see anything that excludes me from what has been granted to this new body of Benedictines, nor can I persuade myself that the singular clemency of Our Father [...] intended to deprive me implicitly of the best prerogatives of membership, that is, to take part in the election of new members when it occurs.[39]

[35]Giuseppe Gaetano Bolletti (1751).

[36]Interestingly, even Giuseppe Veratti's *Osservazioni fisico-mediche intorno alla elettricità* (1748), a book that was arguably the result of their mutual interest in electricity, was dedicated to the Senate of Bologna rather than to Benedict XIV.

[37]AASB, *Registro degli Atti dal 1723 al 1803*, n. 9 (25 August 1745).

[38]*Ibidem* (23 November 1745).

[39]Elio Melli (1960, p. 115) (Bassi to Scarselli, 27 November 1745).

This at least was Bassi's opinion. She asked Scarselli to find out "Our Father's mind" when he created her position though she sincerely hoped that the pope would not disagree with her. She invited Scarselli to send his response to Galeazzi.

Bassi considered Galeazzi a sympathetic supporter. On the whole, this was correct. Galeazzi had made the case for increasing the number of *Benedettini* from 20 to 24, and for including Bassi as the twenty-fifth member. Yet he now found himself confronting a different problem, namely whether "Signora Laura not only may be able [...] to participate in learned exercises, recite dissertations, and enjoy the emoluments that the clemency of Our Holiness has grant to the other Benedictines but also have a role in all the other meetings that regard the election of the president of the academy, and other Benedictines [...]".[40] The majority of *Benedettini* did not believe that these rights were implicit in the papal bull. Like Bassi, they did not wish to interpret the pope's wishes and sought assistance from Scarselli to clear up the situation. Galeazzi was also their chosen intermediary.

The last thing Scarselli wanted to do was to involve the pope. He consulted with Leprotti, writing sympathetically to Bassi of their surprise at this turn of events. "Monsignor Leprotti and I were surprised, not to say disgusted by the incredible difficulties for which the *Motu proprio* certainly did not provide reasonable grounds".[41] They advised everyone to accept their arbitration in order to resolve the situation quickly and discreetly. By mid-December Bassi thanked him for quieting "the doubting minds of our academicians". She did not feel that the discussion had occurred "from any malice towards me" but simply because of the procedural issues that arose in calculating the effect of an additional vote.[42]

Bassi's interpretation of the nature of the disagreement surrounding the definition of her position as *Benedettina* seems to have been her way of coping with an unpleasant reminder that she was never fully invited to participate in the academy's activities. After receiving Scarselli's letter questioning the basis for the *Benedettini*'s decision to exclude Bassi from certain meetings, Galeazzi felt compelled to respond with his own explanation of "the business of Signora Laura" (*l'affare della Sig.a Laura*). He reminded Scarselli of the restrictions placed on Bassi's professorship – "she cannot go to the college unless she is called, and the college only calls her when there is some distinguished doctorate"—and insisted that she had no right to take part in university decisions, including the appointment of other members of the College. Galeazzi then defined the nature of her membership in the Istituto academy:

> The philosophical academy admitted her, then, as it admits all other persons distinguished in learning and rank, but it did not admit her to assume offices, or so that she must come to all the meetings, or meddle in all the academy business. And it does not seem proper to the decency and respectability of her sex to be always in the middle of a meeting of men and obliged to hear all their discussions and quarrels.

---

[40]Biblioteca Universitaria, Bologna (hereafter BUB), ms. 72, I, f. 64r (Galeazzi to Scarselli, Bologna, 27 November 1745).

[41]BCAB, *Scarselli*, I, 9 (Rome, 4 December 1745).

[42]Elio Melli (1960, p. 117) (Bassi to Scarselli, Bologna, 11 December 1745).

Almost all of the Benedettini agreed with this decision, including distinguished senior members such as Bazzani, Beccari, Laurenti, Manfredi, and Peggi. In fact, Galeazzi was pretty sure that the only academicians who disagreed with this view were Bassi and her husband Veratti. He reassured his friends in Rome that this decision was not a product of "any hostility or harshness towards Signora Laura as a person".[43]

We can now see exactly what the problem was. Bassi's appearance at the November 18 meeting reminded her fellow *Benedettini* that, however much they admired and supported her, they simply were not comfortable having her intimately involved in all of their activities, especially the contentious issue of elections. Galeazzi explained it as a matter of moral probity: "It didn't seem decent to them that a lady, even married, ought to always be, as I said, in the middle of the quarrels and discussions of men".[44] He urged Scarselli to consider the justness of the university's decision in 1732 to restrict Bassi's teaching, and hoped that the same limitations would apply to her role in this new category of academician. In the end, the situation was left ambiguous, or as Bassi put it, "neither yes nor no". She had argued forcefully for the importance of merit in such decisions – "to offer justice to those who are deserving".[45]—but in the end she could not entirely negate the fact that her male colleagues would never be comfortable involving her in the kinds of discussions that they were used to having only with each other. Her effect on them was something that neither side could entirely negate.

Having agreed not to have an official ruling on this issue, how did the Istituto handle the question in practice? Bassi attended the regular meetings, both with and without her husband. Yet the first time that two vacant studentships were discussed, neither she nor Veratti attended the meeting of the *Benedettini* on 3 March 1746 *ad eligendum duos alumnos*. Veratti had every right to be present and would be there for 29 December 1747 meeting *ad eligendum alumnum*. In Bassi's case, however, Zanotti tersely noted: "Signora Laura Bassi was not sent a ticket".[46] Bassi did not attend the next three meetings.

The next meeting Bassi attend was on 1 April 1746, when Émilie du Châtelet became the third woman admitted to the academy. The documents do not specify who proposed this distinguished French Newtonian as a candidate. Yet I am inclined to think that Voltaire's witty letter to Bassi in March 1745, in which he imagined himself to be a man between two women in a kind of philosophical *ménage à trois*, expressing the desire to one day have these two women of science meet, did indeed stimulate interest in affiliating Bassi's best living counterpart with the Istituto.[47] Perhaps the uncomfortable discussions about Bassi's status encouraged her to propose Châtelet to

[43]BUB, ms. 72, I, f. 68r-v (Galeazzi to Scarselli, Bologna, 22 December 1745).

[44]*Ibidem*, f. 69r.

[45]Elio Melli (1960, p. 115) (Bassi to Scarselli, 27 November 1745).

[46]AASB, *Registro degli Atti dal 1723 al 1803*, n. 9 (3 March 1746).

[47]Ernesto Masi (1981, p. 170) (Voltaire to Bassi, 1 March 1745). On Châtelet's admission, see AASB, *Registro degli Atti dal 1723 al 1803*, n. 9 (1 April 1746), Mauro De Zan (1987) and Massimo Mazzotti (2008).

join her in creating an imagined community of women scientists within the academy, ceremonial to be sure but nonetheless powerful in the impression that it created of this city as a paradise for women. Possibly she envisioned Châtelet as an ally at a distance given the strength of her reputation and the recent Italian translation of her *Institutions de physique* (1741).[48] Had Châtelet come to Bologna, she would have surely encountered Bassi as the experience of the two other French women admitted to the academy reveals.

When the French artist Marguerite Le Comte, whose natural history illustrations excited great admiration, visited Bologna with her lover Watelet in 1764, she arrived with a letter of introduction to Bassi who subsequently sponsored her admission to the academy.[49] Le Comte became the sixth female member of the Istituto, following the admission of the French poet Anne-Marie du Boccage in 1757. In the case of Boccage—who marveled at the circumstances of her admission since she knew the difference between her own literary interests and the scientific accomplishments of the now deceased Châtelet—Algarotti proposed her membership. At the Istituto Boccage witnessed Bassi's demonstration of Hallerian "experiments on irritability" after attending lectures by the Istituto professors.[50] To differing degrees, each of these women became part of the academy because of the presence of Laura Bassi.

Women admitted to the Bologna Academy of Sciences, 1732–1800

NAME	YEAR	NATIONALITY	SCIENTIFIC SPECIALTY
Laura Bassi	1732	Bolognese	Philosophy; Physics
Faustina Pignatelli	1732	Neapolitan	Mathematics; Physics
Émilie du Châtelet	1746	French	Physics
Maria Gaetana Agnesi	1748	Milanese	Mathematics
Anne-Marie du Boccage	1757	French	Literature
Marguerite Le Comte	1764	French	Natural History/Art
Marie Dalle Donne	1800	Bolognese	Medicine; Obstetrics

4.3 "Signora Laura's Requests" (1776)

In late April 1746, Bassi gave a dissertation to the Istituto, "a Latin discourse on the compression of air".[51] It was the first of thirty-one dissertations she presented between 1746 and 1777, and also the basis of her first publication, *De aëris compressione*,

[48] Émilie du Châtelet (1743).

[49] AASB, *Registro degli Atti dal 1723 al 1803*, n. 14 (6 December 1764).

[50] Anne-Marie du Boccage (1770) (Bologna, 9 June 1757). On the debates surrounding Hallerian experiments in Bologna, see Marta Cavazza (1997).

[51] AASB, *Registro degli Atti dal 1723 al 1803*, n. 9 (28 April 1746).

which appeared in the second volume of *Commentarii* published in 1745.[52] Or we should more accurately describe it as the first publication of her work since Zanotti presented it as a report of the experiments made by *nostra Laura* that led to important tests of the universality of Boyle's law. Following the lead of Galeazzi's earlier work with thermometers, Bassi discovered that she could only replicate Boyle's results on dry rather than humid days because they did not characterize the behavior of vapor under pressure. It is unclear why she was not the author of her own conclusions. Possibly she hoped to offer a more detailed account of this phenomenon, since she continued to do experiments in this area, and acquiesced to Zanotti's eagerness to see some version in print. But it also is in keeping with her relationship to the academy that the secretary presented Bassi's work on her behalf. In 1791 Sebastiano Canterzani would publish a summary of her ongoing work on fluids in Boyle's vacuum, *De immixto fluidis aëre*, by "the much noted Laura Bassi, wife of Veratti, who formerly debated in the Academy". One of the reasons he published the paper was to ensure that her contributions would not be "knowingly neglected".[53] It was the last of four papers to appear in the *Commentarii.*

Every spring Bassi prepared an annual public presentation on different aspects of mathematical and experimental physics. She never failed to meet this requirement of her position as *Benedettina* and always collected her 50 lire. We should contrast this kind of steady production with the more spectacular strategy of the Milanese mathematician Maria Gaetana Agnesi who became a member in June 1748, in the midst of completing the final revisions on her massive, two-volume mathematics textbook, *Instituzioni analitiche.* Agnesi's printer rushed to add "of the Bologna Academy of Sciences" to the frontispiece so that the entire world would know of her affiliation.[54] Agnesi made sure that a copy reached Bassi who warmly thanked her, writing that the Milanese mathematician had fulfilled her own potential and would be immortalized because of this book but also thanked her for increasing "our luster and dignity".[55] Her sincere admiration for this fascinating contemporary, who was in many respects her rival as Italy's leading woman of science, makes evident her pleasure at the growing community of woman affiliated with the Istituto delle Scienze. Bassi indeed saw herself as an inaugural member of a republic of scientific women who met each other, in part, through the Istituto's decision to place them in the ranks of *Accademici Onorarii.*

Bassi, however, would never write such a book. The kind of teaching Agnesi put on paper mirrored the lessons she gave daily in her home for most of her career. Perhaps putting her name on this kind of publication was not especially interesting; she already had her university professorship and by the end of her life she would hold two other professorships at the Collegio Montalto and the Istituto delle Scienze. Perhaps this is a reminder that Bassi's primary goals as an academician were not honorary at all

[52]AASB, Domenico Piani, *Catalogo dei lavori dell'Antica Accademia raccolti sotti I singoli Autori* (nineteenth century); Laura Bassi (1745).

[53]Laura Bassi (1791, quotes on p. 44).

[54]Massimo Mazzotti (2007) and Paula Findlen (2011).

[55]Biblioteca Ambrosiana, Milan, ms. O. 201 sup., c. 10 (Bassi to Agnesi, Bologna, 18 June 1749).

but quite practical in the advancement of her career and reputation. Bassi wanted to be informed about the work of other scientists, while having opportunities to present and, to a lesser degree, publish her research. She witnessed the warm reception of Agnesi's book, not only by her own academy but also by the Paris Academy of Sciences, an episode noted and discussed by the Bologna academicians.[56] She may have also heard about plans for an English translation and implicitly the approval of the British mathematical community. She was certainly well aware of the great success of Châtelet's *Institutions de physique* that, while presented as a series of physics lessons for her son, was nonetheless a sophisticated treatise of Newtonian and Leibnizian natural philosophy. Yet none of this altered her way of proceeding in part, I think, because she was first and foremost a teacher and an experimenter who preferred demonstrable results.

At the height of her career, when the initial difficulties over her membership in the *Benedettini* seemed a distant memory, what did Bassi hope to accomplish? She wanted to be a full-fledged member of an international community of philosophers and experimenters who increasingly defined their relations, in part, through academy memberships, the simultaneous testing and resting of key experiments, and the desire to share and debate their results. Unlike Agnesi who formally withdrew from correspondence shortly after her father's death in 1752, leaving science behind for a life devoted to faith, Bassi cultivated important epistolary relationships with leading foreign experimental philosophers such as Jean-Antoine Nollet and Giambattista Beccaria, former students such as Lazzaro Spallanzani, Felice Fontana, Leopoldo Marc'Antonio Caldani, and less directly Luigi Galvani, and young Italian experimenters such as Alessandro Volta, desirous of her opinion and patronage. It is a sign of her skill at building a scientific network that she persuaded many of them to seek her out on their own.

But what about Bassi's papers? What can we learn from a list? Three dissertations were mathematical in nature; despite the vagueness of the titles, they nonetheless underscore her competency in algebra and analytic geometry. Ten papers dealt with questions of mechanics. In this choice, Bassi did not neglect her Bolognese roots. As the debates about water rights between Ferrara and Bologna raged unabated in the Papal States, diverting both Manfredi brothers from more theoretical pursuits, Bassi continued to research and write about water. *De problemate quodam hydrometrico* and *De problemate quodam mechanico* were the only papers published by Bassi, in her own voice, during her lifetime; they appeared in the fourth volume of the *Commentarii* in 1757.[57] Both demonstrated her mastery of the previous century and a half of research on hydrostatics and motion, and her understanding of the utility of scientific knowledge just as her 1769 *Prodromo d'una serie di sperienze da fare per perfezionare l'arte della tentura* reveals her contribution to the technical advancement of Bologna's silk industry.[58]

---

[56]AASB, *Registro degli Atti dal 1723 al 1803*, n. 10 (21 January 1750).

[57]Laura Bassi (1757). See Cesare Maffioli (1994).

[58]Marta Cavazza (2006, p. 75).

Bassi also continued to pursue optics, which manifested itself not only in her ongoing replication of Newton's prism experiments for students in her home and occasional demonstrations in the Istituto physics cabinet for illustrious visitors, but also in her 1762 paper, *Sopra il vetro islandico*, followed by her 1763 dissertation *Sopra la maniera di correggere nei telescopi l'inconveniente che nasce dalla diversa refrangibilità dei raggi* that she presented twice to the Istituto, once in a private academy and the second time in a public academy open to the entire city and visitors.[59] This paper involved her in a discussion then underway by Alexis Clairaut and Roger Boscovich about understanding the theory behind John Dollond's achromatic lenses which solved the problem of chromatic aberration that even the great Newton had found intractable.[60] Clairaut and Boscovich became members of academy in the 1740s, and Bassi used her correspondence to eventually acquire the kind of new lenses needed to fully assess Clairaut's theory in relationship to the instrument. In this episode, we see clearly how Bassi used the information networks, equipment, and well-appointed library of the Istituto to keep abreast of the latest scientific developments in England and France. In this respect, she was a full-fledged academician.

Bassi's engagement with the most exciting new developments in the physical sciences is especially apparent in her work on electricity and chemistry. Six papers (possibly seven if we hypothesize that her 1764 paper, *Sopra alcuni fenomeni dei fluidi ricevuti nei tubi di diversa materia*, probably dealt in part with electrified tubes) concerned electricity. Two of her last papers, on the nature of fire and fixed air in 1775 and 1776, reveal her keen interest in recent developments in chemistry in the age of Black, Priestley, and Lavoisier. These papers, including the two articles she published and the two summaries of other papers published on her behalf, constitute an invaluable record for understanding the evolution of her research as a physicist interested in the application and refinement of Newtonian natural philosophy in the electrical experiments of Benjamin Franklin, the physiology of Stephen Hales (who became an academy member in June 1757), and the chemistry of Joseph Priestley.

Even a brief discussion of what we can learn about her dissertations provides us with important background to the final episode in Bassi's relationship with the Istituto: her appointment as professor of experimental physics in 1776. Bassi had asked to be considered for this position for several years. Since 1770 Paolo Balbi had replaced Galeazzi and he, in turn, appointed Veratti as his assistant who took over the teaching responsibilities after Balbi became ill in 1772. Many academicians assumed that Veratti would eventually replace him, and Veratti had his own ideas about how to modernize the teaching and instruments necessary for this field. He proposed dividing the professorship into two subjects—experimental physics and electricity—that would correspond with a reorganization of the physics cabinet as well as the teaching program. Canterzani was not convinced that this division would work; moreover, he had found Veratti a difficult collaborator. During Senator Filippo Aldrovandi's discussions with Canterzani and Veratti in the spring of 1776, the idea

[59] AASB, *Registro degli Atti dal 1723 al 1803*, n. 14 (28 April and 14 June 1763).

[60] Gabriella Berti Logan (1999, pp. 513–515).

of appointing Bassi "in the position of first professor of experimental physics [...] leaving her husband, Signor Verati, as her substitute" was born. The possibility of appointing Bassi and Veratti co-professors, with Canterzani and Buonaccorsi as their assistants, was proposed but quickly rejected because it would compound "the spirit of turmoil and division that has been introduced".[61] Instead, Bassi and Canterzani became the first occupants of the two newly defined professorships in experimental and mathematical physics, each with their own assistant. She chose her husband.

Bassi had fairly earned this position after the great success of her private school of experimental physics for almost three decades. The Senate even acknowledged its utility by increasingly her salary on the condition that her school continued to offer students a kind of scientific education they would not otherwise receive.[62] While Bassi's capacity to teach the entire Istituto physics course was widely known—she was surely as competent, if not better qualified than Canterzani to teach mathematical physics—there were nonetheless lingering resentments about her habit of constantly asking for things she should not. Would it ever be possible "to satisfy, if one ever can, the requests of Signora Laura Bassi who, even though she has no right to be admitted among the Institute professors, nonetheless has been asking for a good three years, having hoped more than once for this outcome?" Yet having said this, both Aldrovandi and Canterzani reflected on the reasons why it might be a good idea: "since she is a famous woman known to the entire Republic of Letters, and who truly does great honor to her homeland, therefore it seems that she deserves favorable consideration by the exalted Assunteria."[63] By August 1776, Bassi and Canterzani were doing a complete inventory of the physics cabinet, throwing out old and broken machines to make way for new items to improve the experimental physics courses.[64]

In his 1778 biography Giovanni Fantuzzi, who was also a member of the academy, explained that Bassi's innovative teaching of experimental physics in her home had earned her the Istituto professorship in 1776.[65] There is every reason to agree with this conclusion when we consider the central role she played in making Bologna an important center for experimental physics in the mid-eighteenth century. How did this begrudging admiration for what she had accomplished affect her standing as an academician? If we look at the Istituto records of meetings, we see a very interesting fact regarding Bassi's much discussed voting privileges. During the July 1768 meeting to elect foreign members, including her cousin Spallanzani, both Bassi and Veratti presented his candidacy but also participated in the *scrutinio* of four other candidates.[66] She increasingly seems to have been present during elections of new *Benedettini* and their student assistants (*alunni*) though she did not vote. While the rules never changed—in fact, during the years immediately before and after her

[61] ASB, *Assunteria di Istituto. Diversorum*, b. 15, n. 42 (6 May 1776).

[62] Marta Cavazza (1995, pp. 715–753).

[63] ASB, *Assunteria di Istituto. Diversorum*, b. 15, n. 42 (6 May 1776).

[64] *Ibidem*, b. 10, n. 10 (8 August 1776).

[65] Giovanni Fantuzzi (1778, p. 14).

[66] AASB, *Registri*, n. 15 (9 July 1768).

death, there were detailed discussions about the temporary nature of the twenty-fifth position—in practice the academicians acknowledged that Bassi was an active and valued participant. When she died, so soon and so suddenly after the death of Francesco Maria Zanotti who had been a pillar of this institution, the academicians accompanied her body from the family home to the church of Corpus Domini where she was buried with great pomp and ceremony in her ermine cap and silver laurels. The inscription on her tomb reminded everyone that Bassi had been the glory of the Istituto.[67]

Perhaps this experience of Laura Bassi made them nostalgic for her presence once she was no longer there. In March 1800, Maria Dalle Donne became the new twenty-fifth member, "in entirely the same way that Benedict XIV of glorious memory carried this out with the late, great Dottoressa Signora Laura Bassi".[68] She would be unable to fully repeat Bassi's success as a long-standing professor and researcher. And yet Dalle Donne was in a way Bassi's final success, witnessed by her youngest son Paolo Veratti, a physician and professor who was present for this decision and still hoping to continue the dynasty of experimental physics inaugurated by his mother. Thus, at the dawn of a new century, the Istituto decided to revive the memory of Laura Bassi. It would not repeat this experiment again for many years.

Figures 4.1, 4.2, 4.3, 4.4, 4.5, 4.6, 4.7, 4.8, 4.9, 4.10 and 4.11 show the portrait of Laura Bassi and of men of science of the time; and some miniatures illustrating relevant events of Laura Bassi's life.

[67]Paula Findlen (2018).

[68]ASB, *Assunteria di Istituto. Diversorum*, b. 9, n. 13 (7 July 1800, archiving a note from 31 May).

Fig. 4.1 Laura Bassi (1711–78), professor of physics at the Istituto delle Scienze. Carlo Vandi, oil on canvas, 18th century, Bologna, Museo di Palazzo Poggi

Fig. 4.2 Pope Benedict XIV (1675–1758), born Prospero Lambertini. Attributed to Carlo Vandi, oil on canvas, 18th century, Bologna, Museo di Palazzo Poggi

Fig. 4.3 Jacopo Bartolomeo Beccari (1682–1766), was trained as a physician but later became professor of physics and then chemistry at the Istituto delle Scienze, Bologna, Museo di Palazzo Poggi

Fig. 4.4 Francesco Maria Zanotti (1692–1778), Secretary of the Istituto delle Scienze and of the Accademia. William Keeble, oil on canvas, 18th century, Biblioteca Universitaria di Bologna, inv. 51

Fig. 4.5 Francesco Algarotti (1712–1764), a student at the institute's Accademia delle Scienze, Biblioteca Universitaria di Bologna, inv. 100

Fig. 4.6 Domenico Gusmano Galeazzi (1686–1775), professor of physics at the Istituto delle Scienze. Oil on canvas, 18th century, Bologna, Museo di Palazzo Poggi

Fig. 4.7 Eustachio Manfredi (1674–1739), professor of astronomy at the Istituto delle Scienze. Attributed to Ercole Lelli, white marble, 1739, Bologna, Accademia delle Scienze

Fig. 4.8 *Laura Bassi defends her philosophical theses in public* (17.4.1732). Miniature by Leonardo Sconzani, from Anziani Consoli, Insignia, vol. XIII, c. 94, 1732, Archivio di Stato di Bologna

Fig. 4.9 *Laura Bassi publicly receives the Insignia of Doctor*. Miniature by Leonardo Sconzani, from Anziani Consoli, Insignia, vol. XIII, c. 95, 1732, Archivio di Stato di Bologna

Fig. 4.10 *Laura Bassi's first Lecture at the Archiginnasio*. Miniature by Leonardo Sconzani, from Anziani Consoli, Insignia, vol. XIII, c. 98, 1732, Archivio di Stato di Bologna

Fig. 4.11 *Laura Bassi participates in the discussion at the Public Anatomy* (Carnevale 1734). Miniature by Bernardino Sconzani, Anziani Consoli, Insignia, vol. XIII, c. 105, 1734, Archivio di Stato di Bologna

Acknowledgement Many thanks to Raffaella Simili for inviting me to participate in the Forum Laura Bassi, to Marta Cavazza, Paola Govoni, and Giuliano Pancaldi for making this research a pleasure over many years, and to Massimo Zini for providing me with access to the archive of the Istituto delle Scienze and the benefit of his knowledge of these materials during multiple visits.

References

Francesco, Algarotti, *Non la Lesboa*, in *Rime per la famosa laureazione ed acclamatissima aggregazione al Collegio filosofico della ill.ma ed ecc.ma sig.ra Laura Maria Catterina Bassi*, Bologna, 1732, pp. 23–24

Francesco, Algarotti, *Neutonianismo per le dame*, Naples [Milan], 1737

Annarita Angelini (Editor), *Anatomie accademiche*, vol. 3, *L'Istituto delle Scienze e l'Accademia*, Il Mulino, Bologna, 1993

Laura Bassi, "De aëris compressione", *De bononiensi Instituto atque Accademia Commentarii*, vol. 2, part 1, Bologna, 1745, pp. 347–353

Laura Bassi, "De problemate quodam hydrometrico", *De bononiensi ... Instituto atque Accademia Commentarii*, vol. 4, Bologna, 1757, *Opuscula*, pp. 61–73; eadem "De problemate quodam mechanico", pp. 74–79

Laura Bassi, "De immixto fluidis aëre", *De bononiensi ... Instituto atque Accademia Commentarii*, vol. 7, Bologna, 1791, pp. 44–47

Gabriella Berti Logan, *The Desire to Contribute: An Eighteenth Century Italian Woman of Science*, «American Historical Review», 99, 1994, pp. 785–812

Gabriella Berti Logan, *Italian Women in Science from the Renaissance to the Nineteenth Century*, Ph.D. diss., University of Ottawa, 1999

Paola Bertucci, *Viaggio nel paese delle meraviglie. Scienza e curiosità nell'Italia del Settecento*, Bollati Boringhieri, Torino, 2007, esp. pp. 162–166, 204–211

Giuseppe Gaetano Bolletti, *Dell'origine e de' progressi dell'Instituto delle Scienze di Bologna*, Bologna, 1751, p. 50

Marta Cavazza, *Settecento inquieto. Alle origini dell'Istituto delle Scienze di Bologna*, Il Mulino, Bologna, 1990

Marta Cavazza, *Laura Bassi e il suo gabinetto di fisica sperimentale: realtà e mito*, « Nuncius » , 10, 1995

Marta Cavazza, *La recezione della teoria halleriana nell'Accademia delle Scienze di Bologna*, « Nuncius » , 12, 1997, pp. 359–377

Marta Cavazza, *Les femmes à l'Académie: le cas de Bologne*, in *Académies et sociétés savantes en Europe (1650–1800)*, sous la direction de Daniel-Odon Hurel et Gérard Laudin, Honoré Champion, Paris, 2000, pp. 161–175

Marta Cavazza, *Una donna nella repubblica degli scienziati: Laura Bassi e i suoi colleghi*, in *Scienza a due voci*, a cura di Raffaela Simili, Olschki, Firenze, 2006

Marta Cavazza, *Laura Bassi and Giuseppe Veratti: An Electric Couple during the Enlightenment*, «Contributions to Science» , 5, 2009, pp. 115–128

Marta Cavazza, *Laura Bassi. Donne, genere e scienza nell'Italia del Settecento*, Editrice Bibliografica, Milano, 2020

Beate Ceranski, *"Und sie fürchtet sich vor niemandem": Die Physikerin Laura Bassi*, Campus Verlag, Frankfurt, 1996

Mauro De Zan, *Voltaire e M.me du Châtelet. Membri e corrispondenti dell'Accademia delle Scienze di Bologna*, « Studi e memorie dell'Università di Bologna » , n. s., vol. 6, 1987, pp. 141–158

Mauro De Zan, *L'Accademia delle scienze di Bologna: l'edizione del primo tomo dei* Commentarii *(1731)*, in *Scienza, filosofia e religione tra '600 e '700 in Italia*, a cura di Maria Vittoria Predeval Magrini e T. Boaretti, Franco Angeli, Milano, 1990, pp. 203–259

Anne-Marie du Boccage, *Lettres sur l'Italie*, in Eadem, *Recueil des Œuvres*, Paris, 1770, vol. 3, p. 175
Émilie du Châtelet, *Istituzioni di fisica di Madama la Marchesa Du Chastellet indiritte a suo figliuolo*, Venezia, 1743
Giovanni Fantuzzi, *Elogio della dottoressa Laura Maria Caterina Bassi Verati*, Bologna, 1778
Paula Findlen, *Science as a Career in Enlightenment Italy: The Strategies of Laura Bassi*, «Isis», 84, 1993, pp. 441–469
Paula Findlen, *The Scientists's Body: The Nature of a Woman Philosopher in Enlightenment Italy*, in *The Faces of Nature in Enlightenment Europe*, edited by Gianna Pomata and Lorraine Daston, Berliner Wissenschaftsverlag, Berlin, 2003, pp. 211–236
Paula Findlen, *Women on the Verge of Science: Aristocratic Women and Science in Early Eighteenth-Century Italy*, in *Women, Equality and Enlightenment*, edited by Sarah Knott and Barbara Taylor, Palgrave Press, London, 2005, pp. 265–287
Paula Findlen, *Calculations of Faith: Mathematics, Philosophy, and Sanctity in Eighteenth-Century Italy (New Work on Maria Gaetana Agnesi)*, «Historia mathematica», 38, 2011, pp. 248–291
Paula Findlen, "The Scientist and the Saint: Laura Bassi's Enlightened Catholicism", in *Catholic Women of the Enlightenment*, edited by Ulrich Lehner, Routledge, London, 2018, pp. 114–130
Cesare Maffioli, *Out of Galileo: The Science of Waters, 1628–1718*, Erasmus Publishing, Rotterdam, 1994
Ilario Magnan Campanacci, *La cultura extraccademica: le Manfredi e le Zanotti*, in *Alma mater studiorum. La presenza femminile dal XVIII al XX secolo*, CluEB, Bologna, 1988, pp. 39–67
Ernesto Masi, *Laura Bassi ed il Voltaire*, in *Studi e Ritratti*, Zanichelli Bologna, 1881
Massimo Mazzotti, *Mme Du Châtelet académicienne de Bologne*, in *Émilie Du Châtelet, éclairages et documents nouveaux*, sous la direction de Olivier Courcelle et Ulla Kölving, Centre international d'étude du XVIIIe siècle, Ferney-Voltaire, 2008, pp. 121–126
Massimo Mazzotti, *Newton for Ladies: Gentility, Gender, and Radical Culture*, «British Journal for the History of Science», 37, 2004, p. 120
Massimo Mazzotti, *The World of Maria Gaetana Agnesi, Mathematician of God*, Johns Hopkins University Press, Baltimore, 2007
James E. McClellan III, *Science Reorganized: Scientific Societies in the Eighteenth Century*, Columbia University Press, New York, 1985
Michele Medici, *Elogio di Gaetano Tacconi*, «Memorie dell'Accademia delle Scienze dell'Istituto di Bologna», vol. 2, 1864, p. 214
Elio Melli, *Epistolario di Laura Bassi Veratti*, in *Studi e inediti per il primo centenario dell'Istituto Magistrale Laura Bassi*, STEB, Bologna, 1960
Richard Leonard Rosen, *The Academy of Sciences of the Institute of Bologna, 1690–1804*, PhD diss., Case Western Reserve University, 1971
John Stoye, *Marsigli's Europe, 1680–1730*, Yale University Press, New Haven, 1994
Walter Tega (Editor), *Anatomie accademiche*, vol. 1, *I Commentari dell'Accademia delle Scienze di Bologna*, Il Mulino, Bologna, 1986
Walter Tega (Editor), *Anatomie accademiche*, vol. 2, *L'Enciclopedia scientifica dell'Accademia delle Scienze di Bologna*, Il Mulino, Bologna, 1987
Giuseppe Veratti, *Osservazioni fisico-mediche intorno alla elettricità*, Bologna, 1748

Chapter 5
Giovanni Bianchi: A Sensitive Promoter of the World of Women

Miriam Focaccia

"Know then that in this City as well as the famous Signora Doctor Laura Bassi [...] who is known to the whole world for her Physical-Mathematical studies, there is Signora Anna Manzolini, wife of the painter Signor Giovanni Manzolini who, having been the partner of another Painter Signor Ercole Lelli and celebrated Anatomist, taught Anatomy to this Signora Anna his wife which she now practices by herself, cutting into human cadavers as much as any other expert Anatomist". In 1754 with these words the well known anatomist from Rimini Giovanni Bianchi presented to a Florentine friend "two remarkable persons from the City of Bologna".[1]

Indeed he, a remarkable figure in Eighteenth Century Italian culture, as well as maintaining relations with some of the most illustrious scientists of his age including, to name a few, Albert Haller, Giovanbattista Morgagni, Antonio Vallisneri, Iacopo Bartolomeo Beccari, Domenico Gusmano Galeazzi, Gerard Van Swieten,[2] also had personal relations and exchanges of correspondence with several women who were exponents of the scientific and literary culture of his time. These included, as well as Bassi and Manzolini already mentioned, the mathematician Maria Gaetana Agnesi and the Bolognese noblewoman Laura Bentivoglio Davia.

The Eighteenth Century, as is well known, bore witness to profound transformations that overwhelmed the traditional models of behaviour of the female universe. Just as was happening in France with Madame du Châtelet,[3] Voltaire's muse and

[1] Giovanni Bianchi (1754).

[2] His vast correspondence, kept at the Gambalunga Library in Rimini contains a multitude of famous names. In this regard we refer to: Alessandro Simili (1964, 1965).

[3] Judith P. Zinsser and Julie Candler Hayes (2006).

M. Focaccia (✉)
Museo Storico della Fisica e Centro Studi e Ricerche 'Enrico Fermi', Piazza del Viminale 1, Rome, Italy
e-mail: miriam.focaccia@cref.it

'Enrico Fermi' Historical Museum of Physics and Centre for Study and Research, Rome, Italy

L. Cifarelli and R. Simili (eds.), *Laura Bassi–The World's First Woman Professor in Natural Philosophy*, Springer Biographies,
https://doi.org/10.1007/978-3-030-53962-7_5

translator of Newton's *Principia*, a correspondent of Maupertuis, Euler and Bernoulli amongst others, some women, mainly from aristocratic or upper middle class families, began to rebel against the subordinate roles to which they had been relegated by the patriarchal structure of the society of the time, thus conquering new positions on an educational level. As Paula Findlen has highlighted, women began to "populate science",[4] above all as privileged interlocutors of scientific texts, from the *Entretiens sur la pluralité des mondes* by Bernard le Bovier de Fontenelle to Francesco Algarotti's *Newtonianesimo per le dame* in 1737.[5]

The adroit orchestrator of the female presence within the cultural institutions in Bologna was the cardinal Prospero Lambertini, the future Benedict XIV, who encouraged and advanced the careers of Anna Morandi Manzolini and Laura Bassi Veratti as part of a general project of intellectual renewal. The place chosen to launch this project, which was also meant to succeed in transforming the now somewhat faded European image of the city of Bologna, was the Istituto of the Accademia delle Scienze, founded in 1711 by General Luigi Ferdinando Marsili.

It should be remembered that Bologna, a "fascinating and contradictory" city[6] at the beginning of the Eighteenth Century, despite having been part of the Papal States for at least two centuries, had maintained a sort of continuity with the city's Republic by means of the formula of mixed government, that is to say a blend of papal *monarchy* and people's *republic*. This often gave rise to internal and external compromises which clearly distinguished the political features of the city. Instead, with Lambertini's appointment as archbishop in 1731, a period of harmony between the authority of Rome and Bologna's ruling class began that promoted a productive collaboration, both at a general political level and a local level, between pastoral authority and civil, legatine and senatorial power.[7]

Indeed the archbishop possessed exceptional powers as guarantor and mediator, powers that he would also continue to employ constantly with regard to his city, later as pope.[8]

Due to his education in contact with the men of the late Seventeenth Century intellectual renaissance and with the Roman-Bolognese aristocracy infused with libertinism, tolerant by heartfelt conviction, he shared Bologna's desire for political autonomy, an autonomy considerably more advanced than society in the rest of the Papal States, including the capital. Aware of the problematic nature of the Church's temporal power, he had also developed the conviction that he had to draw closer to civil society, to modern culture and the new science, persuaded that compatibility between them would constitute the must authentic nucleus of Christian doctrine. So Lambertini wanted to use Bologna as the testing ground for his innovating and

[4]Paula Findlen (1995).

[5]Bernard Le Bovier de Fontenelle (1687), Francesco Algarotti (1737).

[6]Alfeo Giacomelli (1979).

[7]Alfeo Giacomelli (1996).

[8]Alfeo Giacomelli (2000).

reforming ambitions, relying on the middle classes and the doctoral class that represented them, with the aim of founding a more balanced and more culturally advanced society.

From this standpoint he also proposed to provide a female model which both Bassi and Morandi fitted perfectly: mothers, but also professionals. Unlike what happened, for example, in Naples with Maria Angela Ardinghelli and Faustina Pignatelli, or in Milan with Maria Gaetana Agnesi,[9] the two Bolognese women, who hailed from a modest social and economic context, lay claim to the importance and "public" usefulness of their expertise, for which they demanded appropriate remuneration.

Certainly Bianchi, linked as he was to the most progressive intellectual milieux of the city, was aware of these inclinations and the impression that emerges from this correspondence is that he tried, albeit politely and with flattering tones, to make use of the fame and network of acquaintances of his female interlocutors as much as possible and in a deceptive way with the aim not only of bringing his own point of view to the centre of attention but also his own person and scientific production, which was rather often involved in harsh debates and fierce criticisms.

Born in Rimini on 3 January 1693, the second of seven children, he was the son of the city's apothecary; his father died unexpectedly in 1701 and the family found itself in distressed economic circumstances and also divided by disputes and squabbles.

Young, brilliant and independent, he began to study with the Jesuits but soon abandoned that school and carried on self taught; "after which he applied himself totally to reading the Historians, Geographers, Botanists, Chemists and other authors of varied erudition. But above all he took delight in Botany"[10] and the Natural Sciences, to which he devoted several writings, including *De conchis minus notis liber* in 1739,[11] in which he described numerous varieties of little known shells of the Adriatic coast, a work that earned him European fame; and also the reprint of Fabio Colonna's precious and now rare *Fitobasano*, adding to the life of the author a brief account of the Accademia dei Lincei and a list of the academicians.[12]

Convinced that Greek was necessary for botanical research he therefore turned to learning the language with total commitment and dedication. At the same time, for four years and under the guidance of Father Giovanni Bernardo Calabro, a doctor of Physics, he continued his studies of that discipline and experimental Philosophy, concentrating in particular on the works of Gassend, Descartes and Newton.[13]

When a literary and scientific Academy was established in Rimini by the city's bishop Giovanni Antonio Davia, the young man became its secretary, proving his own erudition. This was an academy that mirrored the interests and inclinations also present amongst those Bolognese intellectuals who became the messengers of a new rational school based, on the one hand, on experimentalism and, on the other, on questioning the dogmas and the rules on which traditional science was founded.

---

[9]On this question we refer to: Paula Findlen (1995), Raffaella Simili (2008).

[10]Carlo Tonini (1884).

[11]Giovanni Bianchi (1739).

[12]Giovanni Bianchi (1744).

[13]Antonio Montanari (1993).

These very stimuli were the origin of Marsili's idea of an Institute "that taught more for the eyes than for the ears", according to which the natural, physical, chemical and mathematical sciences should dominate research in general, with the conviction that science depended essentially on empirical knowledge.[14] The model proposed tended towards a type of learning that was true but at the same time useful, able to provide answers to the demands of a society, European society between the Seventeenth and Eighteenth Centuries, that was developing and transforming rapidly.

It was within this milieu that Bianchi met the physician Antonio Leprotti and it was on his advice that he decided to take up the study of medicine. In 1717 he moved to Bologna where he stayed from 1717 to 1719, the year in which he graduated in Medicine and Philosophy, according to the custom of the time.

As teachers he had Iacopo Bartolomeo Beccari for experimental Philosophy, Matteo Bazzani for Medicine, Antonio Valsalva for Anatomy, Lelio Trionfetti and Giuseppe Monti for Botany, Eustachio Manfredi and Geminiano Rondelli for Mathematics. So he was in contact with the most progressive and advanced elements of the university teachers of the age, linked to the Istituto delle Scienze, with whom he often developed relationships of friendship and correspondence.

In particular, as far as medicine is concerned, Bazzani and Valsalva, the latter was a pupil of Marcello Malpighi, were the spokesmen of the renewed Bolognese medical school according to which truth derives from observation guided by theory and supported by experiments repeated several times. Thanks to the work of Malpighi himself the Seventeenth Century was the century of microscopic anatomy, of comparative anatomy and biology. The beginning of the anatomy-pathology school also derived from his teaching, given the importance attributed to autoptic confirmation in drawing the links between a clinical picture and the anatomical system. Giovan Battista Morgagni was particularly sensitive to these demands for innovation. In the period of his stay in Bologna, from 1698 to 1707, he studied with some of the teachers who would also teach Bianchi such as Valsalva, Rondelli and Trionfetti.[15]

It was not by chance that after Bologna Bianchi transferred to Padua in order to follow, amongst others, the teaching of Morgagni, who had moved to that city and who had made pathological anatomy an integral part of medicine and an essential premise for its further development.[16]

[14]From this point of view the presence in Bologna of Galileian experimentalism, which from the second half of the Seventeenth Century represented an important phenomenon which influenced the scientific activity of Bolognese scholars, was decisive. Indeed in Bologna, between 1655 and 1656, the first edition appeared of the *Opere di Galileo;* here moreover both Galileo's immediate pupils, and the scientists of later generations who had studied in Pisa and Padua, lived and taught. Between 1629 and 1647, the year of his death, Bonaventura Cavalieri worked in Bologna; as did Geminiano Montanari; Gian Domenico Cassini and Domenico Guglielmini who, together with other young scholars, had gathered around the Accademia della Traccia or Accademia dei Filosofi, and exerted a strong influence on the affairs of the city between the end of the Seventeenth Century and the beginning of the Eighteenth Century.

[15]Giuseppe Ongaro (1988).

[16]Luigi Belloni (1986).

It was thanks to these teachers that Bianchi embraced an ideal of science typical of the Enlightenment, that is to say an aversion towards the dogmas of tradition and trust in the power of science itself to override prejudices and ignorance, together with the concept of medicine as a practical and experimental art. So much so that, in his opinion, to be good physicians it was necessary to have studied Anatomy, Chemistry and experimental Physics. On his return to Rimini, in 1720, he devoted himself to practising medicine and to studying anatomy using cadavers. This latter would be his main passion but also the cause of various difficulties for him.

In 1741 when he was appointed to the chair in Anatomy at the University of Siena, because of the provocative and arrogant personality that characterised him and that from the beginning was stigmatised by his own biographers, he entered into disagreement with his Tuscan colleagues, accusing them, with a very Galilean comparison, of teaching "paper based anatomy", that is to say pure theory. He put forward instead his own way of teaching, based essentially on practical dissection. In reply the other professors accused him of stinking out the hospital and the whole city with his continuous autopsies. After three years he decided to return home where, however, an appointment awaited him which was in no way inferior, that of "primary physician" of the city no less, and he became the most important physician in Rimini, with the rank of noble citizen and an annual salary of two hundred scudi.[17] Following his vast and eclectic interests, he opened a free private school in his home[18] and he made his rich library available to his pupils, together with a museum that contained archaeological, numismatic, botanical and anatomical collections.[19] Amongst his numerous students there was also Laura Bentivoglio Davia.

Now an outstanding figure in the cultural life of Rimini, in 1769 Pope Clement XIV conferred on him the post of honorary papal primary physician, then confirmed by Pius VI, and he became a member of some of the most important scientific and literary academies in Europe including, as well as the academy of Berlin, the Accademia of the Istituto in Bologna.[20]

On this question it should also be remembered that in 1745 he refounded, in his own home, the glorious Accademia dei Lincei that had been founded in Rome by Federico Cesi in 1603 and that had counted amongst its members even Galileo himself. Bianchi, calling himself "Restitutor perpetuus", rewrote its laws[21] and brought together the best scholars establishing, alongside the class of natural sciences, the class of moral and philological sciences.[22]

As we have said, after returning definitively to Rimini after the interval in Siena, as well as devoting himself to teaching in his school—which contemporaries called "Athenaeum"—and to cultural projects such as the restoration of the Lincei, with the

[17] Angelo Fabi (1968).

[18] Giampaolo Giovenardi (1777).

[19] Stefano De Carolis (2003a).

[20] Giampaolo Giovenardi (1777), in the Commentari of the Bologna Institute several of his dissertations on various subjects are present from 1731: see Tega (1986).

[21] «Novelle Letterarie pubblicate in Firenze», VI, 1745, pp. 842–46.

[22] See Carlo Giambelli (1879), Antonio Montanari (2001).

aim of de-provincialising Rimini and returning it to the cultural glories of his youth, Giovanni Bianchi turned his energies to medical practice.

As a physician he always claimed to practice simple and useful medicine and he was a harsh accuser of "those many specious remedies"[23] which in his view did nothing other than prolong disease and more often provoke the patients' death rather than a cure. He was even nicknamed the physician of the desperate, precisely because of his "prodigious" cures in cases that were considered extreme. The introduction of the remedy quinine against ague was due to him, together with the battle against the practice of vesicants, considered a strange and indeed cruel treatment. Certainly he was not always farsighted and progressive, as with regard to the practice of "variolisation", of which he was a harsh critic and opponent, considering "that one should not, and that in conscience one cannot carry out this graft, attempting something doubtful and bringing on a certain illness, which sometimes proves mortal to one who is healthy".[24]

Nor did he limit himself to medical practice: he continued to teach Anatomy and Surgery in the city hospital, not restricting himself to theory but teaching "over the machine itself, continuously opening up Cadavers, or making others open them, so that his Disciples could observe the admirable distribution and connection of all the parts, and the various ways in which such a fine machine is altered or ruined by internal or external causes, and the great mistakes that are made every day by those Physicians or Surgeons who feel revulsion towards opening up human Cadavers in order to learn how to carry out their profession, which cannot likely be done without a good mastery of Anatomy.[25]

His medical works were usually not weighty and for the most part polemical in tone. They ranged from clinical studies to Anatomy, both normal and pathological, and contain interesting indications and observations.

Amongst these the one most worthy of notice is the *Storia medica d'una postema nel lobo destro del cerebello*, in 1751, in which he maintained that the organ in question had direct connections with the nervous system and not crossed over as in the case of the brain. Neurological studies would then prove him right. It was because of this publication, based on the anatomical-pathological observation of the autoptic reports on the "little count" Giambattista Pilastri, that the bitter disagreement we will later discuss would be born.

As we have indicated, in his relations with some women, scientists and educated persons who were his contemporaries, Bianchi showed an attitude of open-mindedness and respect without any kind of prejudice because of their sex.

The letters addressed to him from the noblewoman Laura Bentivoglio Davia, kept in the Gambalunga library in Rimini, are exemplary in this context. Sister-in-law to Bishop Davia and a pupil in Bianchi's school in Rimini in the 1720s, she was

[23]Giampaolo Giovenardi (1777).

[24]Cited in Stefano De Carolis (2004). According to some biographers around 1766 the doctor from Rimini changed his mind regarding inoculation after verifying the good results achieved by physicians.

[25]Giampaolo Giovenardi (1777).

praised by Francesco Maria Zanotti as the "fair Cartesian".[26] Although she defined herself as "a Woman only fit to wield needle and spindle",[27] in her letters she asked Bianchi for explanations of physics phenomena such as "the phenomenon of light",[28] explanations that Bianchi offered her precisely. It also seems that it was to her, as first woman reader, that Bianchi addressed a copy of the *Breve storia di Caterina Vizzani* in 1744 where, in a Boccaccio-esque style and with aims that were more literary than scientific, he analysed the case of androgyny of a woman who lived her whole life as a man.[29]

In 1750 Laura Bentivoglio, curious about some electrical phenomena, submitted to Bianchi's attention a phenomenon "that it seems to me that I have read either in the Anglican Transitions, or in a work of yours [...] This is the phenomenon. A Nobleman amongst the Foremost in Bologna both by birth and Personal Gifts, and a great friend of mine for some time has been forced into a sedentary life by a malady of Urine, with suspicion of the Stone; this misfortune has made him pass from an exorbitant fatness to equal leanness. Talking with him he told me that in the evening, taking off his socks, of which he wears three pairs, and removing them all with one pull, when the Servant separates them one from the other, they spark fire accompanied by a little crackle. The first socks are made of floss, the second of Yarn, the third of Silk"..[30]

A few years before, to be precise in 1746, *Dell'elettricismo, o sia delle forze elettriche de' corpi*, the first treatise on the subject in Italian, had been printed anonymously.[31]

In particular, in Bologna, for some years electrical phenomena and their effects on living organisms had already been at the centre of attention of the Istituto delle Scienze, with crucial attention to their applications in the therapeutic field.

Giuseppe Veratti, Laura Bassi, Tommaso Marini, Iacopo Bartolomeo Beccari shared, between the 1730s and 1740s, a special interest in electrical physics and medicine: electrical experiments and discussions on irritability thrilled the academy

[26]In those same years another noblewoman was praised as the "fair Cartesian": she was Eleonora Barbapiccola from Naples who, in 1722, published a translation into Italian of Descartes' *Principia Philosophiae*. Paula Findlen (2005), Raffaella Simili (2008).

[27]Again Barbapiccola, in the foreword to her translation, with the title "the translator to the reader", a manifesto of women's rights to education and teaching, used a similar expression: "I would hope that you on coming across the title of this book, and seeing that it was the work of a woman, would not send her to Cops, Spindles, and Cloth".

[28]Laura Bentivoglio Davia to Giovanni Bianchi, Bologna, 24 July 1732, *Lettere autografe al Dott. Giovanni Bianchi,* Ms. Gambetti, Biblioteca Civica Gambalunga, Rimini.

[29]Laura Bentivoglio Davia to Giovanni Bianchi, Bologna, 30 November 1744, *Lettere autografe al Dott. Giovanni Bianchi*, cit. It seems that Bentivoglio told the story to Bassi who reacted with great surprise. On this subject and on the editorial vicissitudes and the controversy aroused by this work by Bianchi, we refer to Paula Findlen (2009).

[30]Laura Bentivoglio Davia to Giovanni Bianchi, Bologna, 24 January 1750, *Lettere autografe al Dott. Giovanni Bianchi*, cit. So Laura therefore read the *Philosophical Transactions* of the Royal Society.

[31]We refer in particular to Paola Bertucci (2007).

and the city of Bologna.[32] In 1748 came *Osservazioni fisico-mediche intorno all'elettricità* by Veratti.

Despite their common interests, perhaps because Bentivoglio's philosophical-scientific orientation was from the beginning decidedly Cartesian while Bassi's, after she had emancipated herself from the more traditional teacher Tacconi, had turned towards Newtonism, the former was not only decidedly critical of Laura Bassi's success but also decidedly irritated by the "triumph" that followed her doctorate which she defined, not mincing her words, as "ridiculous".[33] At first, despite the fact that Bassi's success had been orchestrated by the highest levels of Bolognese power,[34] Giovanni Bianchi shared his pupil Bentivoglio's opinion, writing to his old mentor Leprotti: "it seems to me that this girl has been too much courted because if she knows no more philosophy than that she exhibited in that thesis, I do not see that she knows any more than what is known by an infinite number of young men of the same age who have commonly studied, under a common teacher. Since this young woman has claimed to distinguish herself from other women attending to things which that Sex is not in the habit of attending she should also show that she had attended to them in a different way, only expounding theses in noble and useful subjects, as it would have been in my opinion if she had only expounded the theses in pure judgement of René Descartes or Newton, who are the two philosophers famous in our times, and not soil the pages, as she has done for the most part, with monkish and peripatetic slop".[35] He did not hesitate to change his mind after a meeting with that same "female philosopher" in Bologna soon after.[36] Although Bianchi's approval of Bassi was not unconditional, he gave her advice to which disciplines and questions to direct her studies, he became one of her most warm supporters, indeed one of her "patrons", as is shown by their exchange of letters, 17 letters altogether, which began in 1733 and which lasted over a decade.[37]

Bianchi even gloated, again to Leprotti, about the compliance and the modesty of the young Bolognese woman who, according to him, "with great attention and docility was attentive and approved my advice".[38] This is another proof of his arrogant character: indeed in 1733, Laura Bassi had already attained fame and success and was in direct contact with some of the most famous intellectuals of the age!

---

[32] On this question see Walter Bernardi (1992), Marta Cavazza (1997), Marco Bresadola, Marco Piccolino (2003), Miriam Focaccia and Raffaella Simili (2007).

[33] Laura Bentivoglio Davia to Giovanni Bianchi, Bologna, 17 June 1732, *Lettere autografe al Dott. Giovanni Bianchi*, cit. Regarding Bassi's education and the controversies that followed her degree and her university lecture we refer in particular to Beate Ceranski (1996); Paula Findlen (1993); Marta Cavazza (2006).

[34] See: Paula Findlen (1993); Marta Cavazza (2006).

[35] Giovanni Bianchi to Antonio Leprotti, Rimini, 18 May 1732, in Gian Ludovico Masetti Zannini (1979).

[36] Giovanni Bianchi to Antonio Leprotti, Rimini, 12 February 1733, *Ibid*.

[37] The correspondence has been published in full by Beate Ceranski (1994).

[38] Giovanni Bianchi to Antonio Leprotti, Rimini, 12 February 1733, in Gian Ludovico Masetti Zannini (1979).

In his first letter, dated 10 February 1733, Bianchi expressed unhesitatingly esteem and helpfulness towards Laura: "that I here or elsewhere anywhere I might find myself, I shall always live remembering your worth, and I will always recommend it and in every place with everyone, as I have now begun, incessantly [...] I hope that at least the force of my words, the sublime learning, and the high worth of Your Excellency by teaching may arrive at this, that it deserves to pass to the memory of Future generations, so that I could peradventure flatter myself to have satisfied in no small part to that you deserve, and that I rightly have conceived of you".[39] So he hurriedly turned about face, clearly changing sides, perhaps aware of the consequences even at a personal level that supporting Bassi might bring him, thus riding the enthusiasm and expectations of the most modern and progressive part of Bolognese culture that had gambled decisively on the young woman.

In the following letter dated 7 June 1733, Bianchi, putting into play a strategy that was already usual for him, passed off some notes and revisions of Giovanni Crivelli's *Elementi di Fisica,* which in fact was simply a harsh criticism of that text, as advice for the young scholar (who in those same years was perfecting her knowledge of physics and mathematics with teachers such as, for example, Gabriele Manfredi). Apart from the usual compliments to the author, called "most eminent", "most learned" and "excellent", typical of the age, the result was that it was Bianchi himself who was brought to the centre of attention. For the most part he cut out for himself the role of educator of the young, pointing out the "blunders" present in that text and claiming "to benefit the Students of Italy, especially the youngest who by themselves at first sight do not know how to recognise them and avoid them". He then advised Bassi in a rather charming manner to distribute the letter, moreover stressing that would be a sort of privilege for her, so much so that he actually left her: "the total arbiter to make it become as best she pleased".[40] Again in the letter of the following 4 July, despite some perplexity from Bassi, he insisted on this latter point, giving her the freedom to cancel some parts that to her seemed too scornful, as long as those lines of his were made public.[41]

As Beate Ceranski has already acutely analysed, Laura managed to avoid being involved in this disagreement, both because her position within the intellectual circle was not yet consolidated and also because she was aware of Bianchi's difficult character and the numerous diatribes in which he had been involved. And so she promptly replied that she would not make it known to anyone "who receiving it from my hands might doubt that I wished too boldly to meddle in Literary demolishing"[42]: thus avoiding the dangerous trap.

In these letters, as well as the routine deference towards Bassi's learning, Bianchi also undertook personally to help her to improve her knowledge: in particular, as well as sending her copies of his works, he sent her, written in his own hand, a summary of English grammar which, she assures, "I will at once make use of", immediately

[39]Giovanni Bianchi to Laura Bassi, Rimini, 10 February 1733, in Beate Ceranski (1994).

[40]Giovanni Bianchi to Laura Bassi, 7 June 1733, *Ibid.*

[41]Giovanni Bianchi to Laura Bassi, Rimini, 4 July 1733, *Ibid.*

[42]Laura Bassi to Giovanni Bianchi, Bologna, 11 July 1733, *Ibid.*

after having concluded her studies of Algebra, "and if despite the great clarity of it I happened upon some difficulties due to a shortcoming of my intelligence, I will not fail to beg you to expand on them to me".[43]

He would return again to this question years later in more than one missive, stressing how much importance he attributed to the question of language, so much so that, again in 1738, he enquired about his pupil's progress in this field, "so that I sent you the Dictionary printed by Boyer, and a manuscript Summary",[44] insisting and repeatedly emphasizing its importance for the Newtonian studies that Laura was delving into. From this one of the many facets of the learned Bianchi emerges: the enthusiast and profound expert of foreign languages, as well as the classic languages, as well as his modernity in stressing the advantage of being able to understand the works of foreign authors in the original.

On 19 April 1738 Bianchi sent Laura Bassi his congratulations for her wedding. The justifications and arguments that arose around the decision of the Bolognese "Minerva" to marry are well known.[45] Yet it seems that Bianchi clearly distanced himself from those critics who saw a peril for Laura's studies in this marriage: " [...] I now sincerely am delighted for Your Illustriousness for this happy event, wishing you for a long series of years both a happy and carefree life [...] From this I say that all should see that I am not one of those rigid men who interpret matters malevolently and who might almost condemn this marriage of Yours, as a sign of abandoning your studies, because rather I always believe that this agreeable association will serve you as a comfort to ever more happily advance along the path you have engaged in, and more so, since I see that in this match not only you have chosen a learned and talented person, but also timely and suitable [...]".[46] In this case, her prompt reply expresses all Laura's gratitude for his support for her marriage: "As for the congratulations, that you were so kind to send me on my marriage, I am infinitely grateful"; she also reaffirmed the reasons that had pushed her towards this decision: "[...] but finally my domestic circumstances led me to change opinion, and to cling to this decision, as a result of which I am greatly pleased to have met in Your Excellency such a learned appraiser of things, such as you are, because you are unable to condemn it as a total separation from those studies, that I am obliged to profess, and to which instead I claim to be able in this way to attend more quietly, in more freedom, and so I have chosen a person who walks the same road of Letters, and who from long experience I was sure would not turn me aside from it".[47]

---

[43]Laura Bassi to Giovanni Bianchi, Bologna, 24 July 1733, *Ibid*.

[44]Giovanni Bianchi to Laura Bassi, Rimini, 19 April 1738, *Ibid*.

[45]We refer you in particular to Paula Findlen (1993).

[46]Giovanni Bianchi to Laura Bassi, Rimini, Rimini, 19 April 1738, in Beate Ceranski (1994).

[47]Laura Bassi to Giovanni Bianchi, Bologna 26 April 1738, *Ibid*. However on the question of the marriage, which was certainly a delicate matter precisely because of the diatribes it gave rise to, a small misunderstanding occurred between the two scientists due to a sarcastic but playful quip by Bianchi which considerably annoyed the new bride. Bianchi hurried to reply that his words were spoken "as a pleasantry and almost as said in jest", and the incident was not spoken of again. In particular we refer you to Beate Ceranski's article already cited.

If Bianchi took an interest in Laura's progress in her studies and supported her in her entry into the scientific community, also by interceding with other scientists, for example with Leprotti,[48] it is also true that through her he arranged for his own works to be circulated in their community. Indeed, as well as continuously proffering greetings to colleagues, he arranged through her for various of his works to arrive in Bologna, such as, for example, a text of his "on the Phenomenon of the latest Aurora Borealis, that some might quite rightly say could have been a great torch in the Sky fostered by your blessed Nuptials. Now on this phenomenon, as you know, various reports have been issued [...] all done in fact according to the most exquisite diligence of Astronomy but in none of them assigning any reason for the Phenomenon, so many wondered at this, and many have urged me to say something on the question. [...] So I have ventured to present these Explanations to you [...]".[49] Laura immediately asked to be able to divulge it to other "friends" and obviously Bianchi gave his permission,[50] also suggesting to her to communicate it to his old teacher of Botany at the university, Giuseppe Monti, and to the latter's son Gaetàno.

It is worth pointing out that in a later letter, the same tactic is however repeated in reverse: this time Bianchi sent to Bassi via Monti "a copy of my recently printed Booklet that deals with these things of ours which I think are less known, and which has an essay at the end on the Flow and Ebbing of our sea",[51] in other words a copy of *De conchis minus*, asking for her opinion. It is clear that by then Laura had consolidated fame and honours and thus accrued her own scientific independence, so that now it was the mature and celebrated "patron" who asked her to "write to me sincerely your sage thoughts, that I venerate greatly".[52]

A few years later it seems that Bianchi, in 1744, started writing again to Laura, no longer a young scholar in need of guidance but a mature "reader" and academic, extremely famous throughout the Republic of Letters. In her reply she lingered over a very "elegant, really curious and truly unique Story" attached to the letter received from Rimini. It was the *Breve Storia di Caterina Vizzani*, which she already knew about since: "Such a strange episode has been widely talked of, and I heard of it independently from the most worthy Signora Maria Davia one evening, who favoured me with it, we knew enough to understand the mad eccentricity of this woman, who chose to live with trials and die violently for a love totally contrary to that which naturally in women usually reigns".[53] Apart from the interest aroused in many by this case of sexual inversion, she was particularly enthusiastic about the anatomical observations that Bianchi attached to the young woman's autopsy "and especially

---

[48]From this point of view the letters between the two scientists from February to April 1733 are interesting. In them Bianchi actually asked his colleague, the personal physician of Cardinal Davia, for an explanation of the hostile attitude shown towards Bassi both by his noble patient and the latter's sister-in-law, the well known Marchioness Bentivoglio Davia. On this question see: Gian Ludovico Masetti Zannini (1979).

[49]Giovanni Bianchi to Laura Bassi, Rimini, 19 April 1738, in Beate Ceranski (1994).

[50]Giovanni Bianchi to Laura Bassi, Rimini, 3 May 1738, *Ibid.*

[51]Giovanni Bianchi and Laura Bassi, Rimini, 27 June 1739, *Ibid.*

[52]*Ibid.*

[53]Laura Bassi to Giovanni Bianchi, Bologna, 28 October 1744, *Ibid.*

the observations you made on the cystohepathic ducts [...]",[54] thus recalling another disagreement that had involved Bianchi and Giuseppe Pozzi, against his namesake Giovanni Battista Bianchi of Turin and Gaetano Tacconi, Laura's former teacher. The controversy was settled by the famous anatomist Domenico Galeazzi with his memoir *De cystis falleae ductibus* printed in the Istituto's Commentarii, in which the experimental results that refuted the existence of the hepatocystic canals subject of the dispute were reported and thus proved the Bianchi from Rimini to be correct.[55]

Soon after Bianchi wrote to Bassi again in order to make a letter that had appeared in his defence as widely known as possible[56]: it was *Simonis Cosmopolitae epistola apologetica pro Iano Planco ad anonymum Bononiensem*, in fact a work he himself had written, compiled in 1745 in reply to a leaflet particularly hostile towards him.[57]

Evidently he cared a great deal to have his own version against such attacks distributed, so much so, as we shall see, he took care also to send a further copy to another "girl prodigy" of the Eighteenth Century scientific milieu with whom he exchanged letters: the mathematician Maria Gaetana Agnesi, then famous for having published, in 1738, a collection of 191 *Propositiones philosophicae* in which the mathematical sciences were presented as the only ones that allowed certain knowledge. In the four letters written by the young mathematician to the now famous physician from Rimini, which spanned from 1741 to 1744, the traditional roles already seen between Bianchi and Bassi were respected. He sent the young woman copies of his works, specifically *De conchis minus notis*, while Agnesi expressed "most lively thanks " for the gift defined as "precious" and was gratified finally to have some samples of Bianchi's erudition, of which she had indeed heard separately already, and so she promised "At once, when I have gone to Milan from the Villa I will do myself the honour of sending you some copies of my Propositions of philosophy".[58]

In the following letter, dated 10 January 1742, after congratulating Bianchi for his appointment to the chair of Anatomy at Siena and after informing him that she had sent two copies of his book to Venice, Agnesi enlarged upon scientific matters, or rather she informed him that she had changed opinion on two questions presented in the *Propositiones*: "on the figure of our Earth, having seen the latest observations made by the Paris Academicians at the Polar Circle and totally contrary to those that had been made by Cassini and others, on which my first opinion rested"[59]; and on the theory of the aurora borealis, "which as you will observe contains a totally new

---

[54] *Ibid.*

[55] On this question we refer you to Valeria Paola Babini (1987).

[56] Giovanni Bianchi to Laura Bassi, Rimini 23 October 1745, in Beate Ceranski (1994).

[57] The leaflet was by Girolamo del Buono, *In Ioannis Blanci seu Jani Planci Ariminensi Vitam animadversiones anonymo Bononiensi auctore*, which appeared following the biography that was anonymous but attributed to Bianchi himself, more than anything a work of self defence, published in 1742 in the work by Giovanni Lami *Memorabilia Italorum eruditione praestantium* and which set off a series of offensive publications against Bianchi himself, above all after the hostile relations established amongst academics at Siena.

[58] Maria Gaetana Agnesi to Giovanni Bianchi, Valera (Milan), 9 November 1741, in Gino Arrighi (1971).

[59] *Ibid.*

doctrine (although later it has been followed by others) it should be observed that only two reflections of the sun's rays are sufficient to account for all the properties of this celebrated meteor although in the mentioned propositions it is said that three reflections are necessary for the same effect". Although Bianchi had already written several essays on this subject, Agnesi did not refer to them.[60] Some years later, to be precise in 1748, she achieved international fame with the publication of the two volumes of *Instituzioni analitiche per uso della gioventù italiana.* This would lead her, amongst other things, to the office of honorary reader of Mathematics at the University of Bologna, once again at the desire of Benedict XIV: she accepted, but she never carried out her role. In 1752, on her father's death, she abandoned scientific activity to devote herself to spiritual and charitable works, following a vocation already expressed before.[61]

In the last letter, dated 8 December 1744, Maria Gaetana referred to a "vindication made in your favour" that Bianchi sent to her: probably it was again the above mentioned letter signed Simone Cosmopolita.

Again it emerges how Bianchi made use of his female "contacts" to disseminate and defend his own work, which was often, as we have said, at the centre of intense and pungent controversy.

His relationship with Anna Morandi Manzolini, the renowned Bolognese waxwork modeller, who with her husband had given birth to one of the most spectacular productions of anatomical waxworks, was completely different.[62]

Like Bassi, Anna was a protégée of Benedict XIV and in 1756 had received from the Bolognese Senate, after a request addressed directly to the Pope, the appointment as wax modeller at the University's chair in Anatomy, with an initial fee of three hundred Bolognese Lire and without a requirement to give lessons in the University.[63]

[60]His considerations on the aurora borealis, considered of limited scientific value, were published in *Raccolta d'opuscoli scientifici e filologici,* XVII, 1737, pp. 97–105 and 107–117; *Ibid*, XXI, 1740, pp. 185–203. We refer you to Angelo Fabi (1968).

[61]Regarding Maria Gaetana Agnesi see in particular: Antonio Francesco Frisi (1799); Massimo Mazzotti (2006, 2007); Rebecca Messbarger, Paula Findlen (2005).

[62]Regarding Anna Morandi we refer you in particular to: Rebecca Messbarger (2001), Miriam Focaccia (2008), Rebecca Messbarger (2010).

[63]Again like Bassi, Morandi also became one of the "marvels" that the visitors went to admire during their visits to Bologna, also attending the lessons of anatomy that she gave in her own home. When she died, on 9 July 1774, she was buried in the church of San Procolo in Bologna with a solemn funeral as befitted an illustrious citizen. We refer in particular to: Miriam Focaccia (2008).

Anna's creations fully answered the precise requirements of the Bolognese anatomical-surgical school, requirements that would involve Luigi Galvani as a particularly attentive interlocutor, who would employ this new configuration of "anatomy of parts" in the field of his studies on animal electricity. By now an anatomy was turned to that was not merely morphological but physiological, according to a mechanicistic concept of the human body, and a connection had also been forged between Anatomy and clinical practice, with the aim of guaranteeing clear and linear notions to those who had to attend to the sick. In the Bolognese waxworker's preparations the anatomical constitution was therefore extremely modernised at the epistemological level both for the brain-nerves-muscles circuit, and for the novel emphasis given to the sense organs and, in particular, to the faculty of perception.

In her handiwork that "anatomy of parts" is illustrated, concentrated on the organs and inner parts of the human body. However with regard to this anatomy she was not content merely to discover the simple position and structure of the organs but, as Rebecca Messbarger has demonstrated, fixed her attention on their dynamic function. It was a concept of anatomy that perfectly mirrored that of Bianchi, for whom "anatomy of parts" is also spoken of in a mechanicistic conception of the human body.

Not by chance Bianchi showed deep admiration for this "prodigious" woman, while at the same time he involved her, together with her husband, in a dispute he had with three physicians from Cesena, Angelo, Carlo Antonio and Giuseppe Serra, regarding the clinical case of the young Count Pilastri. So although Anna was already known to Bianchi, who indeed had praised her the year before in the pages of the *Novelle fiorentine*, he nevertheless accused her of not being capable of recognising decay in a bone.[64]

Indeed from two letters kept in the Gambalunga library in Rimini, it can be inferred that the dispute was centred on the analysis of a bone, in particular in the case of those who had been "Frenchified", in other words suffering from "French disease" (syphilis), which was carious in the Serra brothers' opinion but healthy according to Bianchi. Moreover the Serras claimed that the bone in question belonged to the "little count" Pilastri, while Bianchi maintained the contrary since he had personally examined the young man's bones and found no sign of decay. At this point the Serras asked for an opinion from the best known physicians of the Bolognese circle "and all of them together concurred in the same sentiment, that is to say that the said fragment of bone was decayed or holed; after which Doctor Serra, who was wholly unknown to us, was brought home by their Excellencies Doctors Galeazzi and Laghi due to the solicitude towards them from Doctors Beccari and Molinelli, so that we could observe again [...]".[65]

[64] Stefano De Carolis (2003b).

[65] Anna Morandi and Giovanni Manzolini to Giovanni Bianchi, Bologna, 15 April 1755, *Lettere autografe al Dottor Giovanni Bianchi*, cit.

The Manzolinis were involved in this controversy indirectly since they were only consulted later. The tone and style of the letters were typical of Anna Morandi. As indeed was her custom she clung firmly to her position even going against Bianchi's opinion and answering his criticisms frankly: "nor are we in any way aggrieved that Your Excellency has said that since we are not surgeons we are not capable of judging the decay of a bone, which has happened because Your Excellency is not well informed that over many years we have dissected hundreds and hundreds of cadavers, setting aside their bones to the benefit of our preparations and as a result we have had great opportunity to discover decay only too well [...]".[66]

Not only that but she elegantly resolved the question in the following terms: "On the question whether to believe that the Bone brought to Bologna by Doctor Serra and that he then had sketched, really belongs to the young Count Pilastri, frankly we maintain to anyone that it is not his, precisely for this reason, because it is decayed; while the Bone of the young Count Pilastri having been examined by Your Excellency, and by the surgeon who makes known to me that at the time the dissection of the Cadaver it was not found by Your Excellency to be in any way decayed; and since we most greatly esteem your expertise on these grounds, we say that it is not his; and be certain that in no respect will we say differently, nothing pressing us with regard to Doctor Serra, but instead the honour and the esteem in which we hold Your Excellency to whom in every respect we profess ourselves to be extremely indebted, in mark of which we shall always show ourselves in deed, as we glory in respectfully proclaiming".[67]

If on the one hand this correspondence shows Bianchi to be cultured and eclectic, even if eccentric, bizarre and impulsive, on the other hand it demonstrates that he had a real inclination "to the feminine" and that they were not isolated episodes, even though he used those exchanges of ideas skilfully and deceptively, spreading his works and opinions via those special female interlocutors. On this point it should be stressed that Bassi herself, now conscious of her own institutional role, made shrewd and strategic choices in advancing her own career precisely with regard to personages of the calibre of the Riminese physician.

Figures 5.1, 5.2, 5.3, and 5.4 show the portraits of Anna Morandi, Maria Gaetana Agnesi, Giovanni Bianchi, and an image of the ancient Bologna University.

[66] *Ibid*, Bologna, 24 May 1755.

[67] *Ibid*.

Fig. 5.1 Attention, care and elegance distinguish Anna Morandi's self-portrait. She depicted herself in extremely feminine garb—note the lavish clothing and jewellery—while intently examining the finely worked and intricate cerebral structure. Bologna, Museo di Palazzo Poggi

Fig. 5.2 Portrait of Maria Gaetana Agnesi

Fig. 5.3 Portrait of Ianus Plancus (Giovanni Bianchi), from Alessandro Simili, «Minerva Medica», LVI, 14, 1965, pp. 1–43

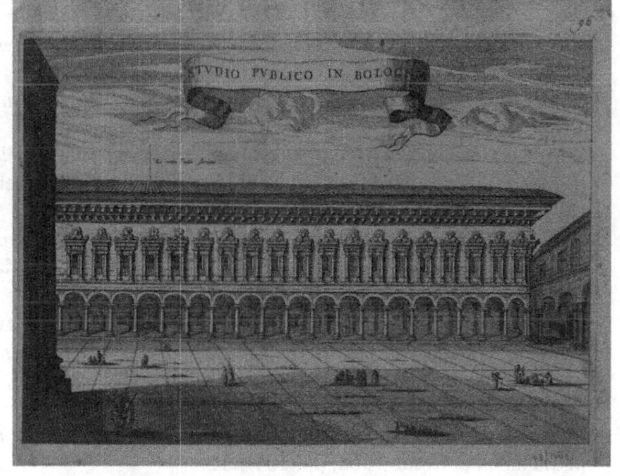

Fig. 5.4 Engraving of the Archiginnasio Palace, the ancient Bologna University centre, Bologna, Biblioteca Comunale dell'Archiginnasio, Cart. Gozzadini, 3-146/1

References

Francesco Algarotti, *Il Newtonianismo per le dame, ovvero dialoghi sopra la luce e i colori,* Naples 1737.

Gino Arrighi, *Incontri di Maria G. Agnesi con Jano Planco. Quattro lettere inedite della scienziata milanese,* «Rendiconti Scientifici dell'Istituto Lombardo», 105, 1971, pp. 681–686.

Valeria Paola Babini, *Anatomica, Medica, Chirurgica,* in *Anatomie Accademiche. L'Enciclopedia scientifica dell'Accademia delle Scienze di Bologna,* II, edited by Walter Tega, Il Mulino, Bologna 1987, pp. 65–67.

Luigi Belloni, *Il pensiero anatomo-clinico del Morgagni e la sua derivazione galileiana,* in *De sedibus et causis. Morgagni nel centenario,* edited by Vincenzo Cappelletti, Federico Di Trocchio, Istituto dell'Enciclopedia Italiana, Rome 1986, pp. 339–345.

Walter Bernardi, *I fluidi della vita:alle origini della controversia sull'elettricità animale,* Olschki Florence 1992.

Paola Bertucci, *Viaggio nel paese delle meravglie. Scienza e curiosità nell'Italia del Settecento,* Bollati Boringhieri,Turin 2007.

Giovanni Bianchi, *De conchis minus notis liber, cui accessit specimen aestus recipcoci Maris Superi ad littus portumque Arimini,* Venetiis 1739.

Giovanni Bianchi, *Fabj Columnae Lyncei Phitobasanos cui accessit vita Fabj et Lynceorum notitia adnotationesque in Phytobasanon Iano Planco Ariminensi auctore,* Florentiae, 1744.

Giovanni Bianchi, *Lettera del Signor Dottor Giovanni Bianchi di Rimini scritta ad un suo amico di Firenze,* « Novelle Letterarie pubblicate in Firenze » , XV, 1754, pp. 708–11.

Marco Bresadola, Marco Piccolino, *Rane, torpedini e scintille. Galvani, Volta e l'lettricità animale,* Bollati Boringhieri, Turin 2003, pp. 83–128.

Marta Cavazza, *Una donna nella Repubblica degli scienziati. Laura Bassi e i suoi colleghi,* in *Scienza a due voci,* edited by R. Simili, Firenze Olschki 2006, pp. 61–85.

Marta Cavazza, *La ricezione della teoria halleriana dell'irritabilità nell'Accademia delle Scienze di Bologna,* «Nuncius», XII, 1997, fasc. 2, pp. 359–377.

Beate Ceranski, *Und sie fuřchtet sich vor niemandem: die Physikerin Laura Bassi (1711–1778),* Campus, Frankfurt 1996.

Beate Ceranski, *Il carteggio fra Giovanni Bianchi e Laura Bassi, 1733–1745,*«Nuncius», IX, 1994, 1, pp. 207–231.

Stefano De Carolis, *Iano Planco medico e scienziato,* in *Atti della seconda giornata amaduzziana (I parte)* (8 April 2001), edited by Giancarlo Donati, Accademia dei Filopatridi, Savignano sul Rubicone 2003a, pp. 5–12.

Stefano De Carolis, *Ripicche e polemiche fra medici nel Settecento: Giovanni Bianchi e il caso clinico del «contino» Pilastri,* «Il Bollettino dell'Ordine dei Medici Chirurghi e degli Odontoiatri della Provincia di Rimini», IV, 2003b, n. 2, pp. 11–19.

Stefano De Carolis, *Chi crede inocularsi si inoculi, chi vuole disinocularsi si disinoculi. Giovanni Bianchi, Francesco Roncalli Parolini e la polemica sull'innesto del vaiolo,* in *Il vaiolo e la vaccinazione in ltalia,* edited by Antonio Tegarelli, Anna Piro, Walter Pasini, vol. II, CNR-WHO, Rome 2004, pp. 621–637.

Angelo Fabi, *Giovanni Bianchi,* in *Dizionario biografico degli italiani,* 10, Istituto della Enciclopedia italiana, Rome 1968, pp. 104–112.

Paula Findlen, *Science as a Career in Enlightenment Italy: The Strategies of Laura Bassi,* «Isis», 84.3,1993, pp. 441–469.

Paula Findlen, *Translating the New Science: Women and the Circulation of Knowledge in Enlightenment Italy,* «Configurations». 3.2, 1995, pp. 167–206.

Paula Findlen, *Giuseppa Eleonora Barbapiccola,* in Rebecca Messbarger, Paula Findlen, *The Contest for Knowledge. Debates over Women's Learning in Eighteenth-Century Italy,* The University of Chicago Press, Chicago & London 2005, pp. 35–66.

Paula Findlen, *Medicine, pornography and culture* in *Italy's eighteenth century: gender and culture in the age of the Grand Tour*, edited by Paula Findlen,Wendy Wassyng Roworth, Catherine M. Sama, Stanford University Press, Stanford, 2009, pp. 216–250.

Miriam Focaccia, Raffaella Simili, *Luigi Galvani Physician, Surgeon, Physicist: from Animal Electicity to Electro-Physiology*, in *Brain, Mind and Medicine: Essays in Eighteenth Century Neuroscience*, edited by Harry Whitaker, C.U.M. Smith, Stanley Finger, Springer, New York 2007, pp. 145–158.

Miriam Focaccia, *Anna Morandi Manzolini. Una donna fra arte e scienza. Immagini, documenti, repertorio anatomico*, Olschki, Florence 2008.

Antonio Francesco Frisi, *Elogio storico di D.a Maria Gaetana Agnesi milanese*, Galeazzi, Milano 1799.

Alfeo Giacomelli, *L'età lambertiniana, in Carlo Grassi. e le riforme bolognesi del Settecento*, «Quaderni culturali bolognesi», III, n. 10, 1979, p. 5.

Alfeo Giacomelli, *Famiglie nobiliari e potere nella Bologna settecentesca*, in *I "Giacobini" nelle legazioni. Gli anni napoleonici a Bologna e Ravenna*, Proceedings of the conference edited by Angelo Varni (Bologna, 13–15 November 1996; Ravenna 21–22 November 1996), Costa, Bologna-Ravenna 1996.

Alfeo Giacomelli, *La Chiesa di Bologna e l'Europa durante l'arcivescovado del cardinal Vincenzo Malvezzi*, in *La Chiesa di Bologna e la cultura europea*, Proceedings of the conference (Bologna 1–2 December 2000), Giorgio Barghigiani Editore, Bologna 2000.

Carlo Giambelli, *L'Accademia dei Lincei*, «Nuova Antologia», s. II, XIV, 1879, pp. 125–151.

Giampaolo Giovenardi, *Orazion funerale in lode di Monsig. Giovanni Bianchi nobile riminese*, Simone Occhi, Venice 1777, pp. XXVII–XXXV.

Bernard Le Bovier de Fontenelle, *Entretiens sur la pluralite des mondes par l'autheur des Dialogues des morts*, Mortier, Amsterdam 1687.

Gian Ludovico Masetti Zannini, *Laura Bassi (1711–1718). Testimonianze e carteggi inediti*, «Strenna Storica bolognese»,1979.

Massimo Mazzotti, *Scienza, fede e carità. Il cattolicesimo illuminato di Maria Gaetana Agnesi*, in *Scienza* a *due voci*, edited by R. Simili, Firenze Olschki 2006, pp. 13–37

Rebecca Messbarger, *Waxing Poetic: Anna Morandi Manzolini's Anatomical Sculptures*, «Configurations», 9/1, 2001, pp. 65–97.

Rebecca Messbarger, *The Lady Anatomist: The life and Work of Anna Morandi Manzolini*, University of Chicago Press, Chicago 2010.

Antonio Montanari, *Notizie inedite su Iano Planco*, Rimini 1993, pp. 1–16.

Antonio Montanari, *Tra erudizione e nuova scienza. I Lincei riminesi di Giovanni Bianchi (1745)*, «Studi Romagnoli», LII, 2001, pp. 401–492.

Giuseppe Ongaro, *Morgagni a Bologna*, in *Rapporti tra le Università di Padova e Bologna*, Editoriale LINT, Padua 1988, pp. 255–287.

Alessandro Simili, «Minerva Medica», LVI, 14, 1965, pp. 1–43.

Alessandro Simili, *Carteggio inedito di illustri bolognesi con Giovamni. Bianchi riminese*, Azzoguidi, Bologna 1964, *Carteggio inedito di Alberto, Haller con Giovanni. Bianchi (Jano Planco)*, «Minerva medica», LVI, 14, 1965, pp. 1–43.

Raffaella Simili, *In punta di penna. Donne di scienza e di cultura fra cosmopolitismo e intimità meridionale*, in *La scienza nel mezzogiorno dopo l'Unitià d'Italia*, Rubettino, Naples 2008, pp. 27–89.

Walter Tega (Editor), *Anatomie Accademiche. I Commentari dell'Accademia delle Scienze di Bologna*, I, Il Mulino, Bologna 1986.

Carlo Tonini, *La coltura letteraria e scientifica in Rimini. Dal secolo XIV ai Primordi del XIX*, Rimini 1884, p. 233.

Judith P. Zinser and Julie Candler Hayes (Editors), *Emilie du Châtelet: rewriting Enlightenment philosophy and science*, Voltaire Foundation, Oxford 2006.

Chapter 6
The Bassi-Veratti Home Laboratory

Marta Cavazza

6.1 "Photographed" and Vanished

Paolo Veratti was the only one of Giuseppe Veratti and Laura Bassi's children to have followed in his parents' footsteps. In 1818, he decided to sell the scientific instrument collection or *gabinetto* that had belonged to his parents and that he himself had used for years, to count Carlo Filippo Aldrovandi Marescotti.[1] The five cupboards containing the substantial "provision of mechanisms for use in experimental physics" (as described in C. F. Aldrovandi's will[2]) that had belonged to the Veratti couple were located in the villa that the Aldrovandi family owned in Camaldoli, a village a few kilometres outside Bologna's city walls. There were over 250 items comprising this collection, every one meticulously described in an inventory dated 1820. The title of the document (*Inventario delle macchine componenti il Gabinetto una volta della fù Sig.ra Laura Bassi-Veratti, ora di ragione del N.U. Sig. Cav. E Conte Carlo Filippo Aldrovandi Mariscotti*),[3] does not make any mention of Giuseppe Veratti although he had undoubtedly taken part in purchasing and using the instruments. Certainly, sources from as early as the eighteenth century identify the *gabinetto* as primarily Bassi's property and the setting for her scientific work and teaching. In 1820, however, this reference to only Bassi assumes special significance. Carlo Filippo was eccentric, a longstanding Jacobin and representative of the Napoleonic government as well

[1]Marta Cavazza (1995a), pp. 715–733. The text of the inventory is provided in the Appendix (pp. 740–753). For the troubled career of Paolo Veratti (1753–1831), first at the *Istituto delle Scienze*, then in the Napoleonic University, see Serafino Mazzetti (1847), p. 158.

[2]Marta Cavazza (1995a), p. 736.

[3]Archivio di Stato di Bologna (from now on ASB), Archivio Aldrovandi Marescotti, b. 430, now reproduced, Marta Cavazza (1995a), pp. 741–743.

M. Cavazza (✉)
Department of Science Education, University of Bologna, Via Filippo Re 6, Bologna, Italy
e-mail: marta.cavazza@unibo.it

L. Cifarelli and R. Simili (eds.), *Laura Bassi–The World's First Woman Professor in Natural Philosophy*, Springer Biographies,
https://doi.org/10.1007/978-3-030-53962-7_6

as a physics and chemistry enthusiast, entrepreneur, philanthropist and patron. He intended to use the collection of instruments he had bought to personally provide a scientific education to the young daughter of one of his servants, Angiola Campeggi, on the basis of a method that he himself had devised. In his opinion, the girl possessed "the rare qualities and excellent disposition" needed to master the sciences.[4] Count Aldrovandi was fascinated by the myth of Laura Bassi, the "*dottoressa*" who in the previous century had made the name of her hometown Bologna famous as the only city in Europe where a woman had made the kind of achievements which everywhere else were reserved to men: not only a doctoral degree, but a paid lectureship at the University, a place in the *Accademia delle Scienze* supported by Pope Benedict XIV, and a position as professor of experimental physics at the *Istituto delle Scienze*.[5] Like other aspiring "Pygmalions" of the time, he hoped to earn the merit and glory of recreating the extraordinary accomplishment of "bringing back" Laura Bassi.[6] This expression, referring to eleven-year-old Maria Dalle Donne, appears in a letter written by Luigi Rodati in 1789. Driven by these ambitions, the count did not hesitate to incur debt in order to purchase the (quite expensive) *gabinetto* and will it to his protégée together with a pension. His death in 1823 interrupted his plans. The Aldrovandi heirs tried to prevent his bequests from being carried out and, after complicated judicial proceedings, Angiola Campeggi and her relatives had to settle for a fairly modest disbursement. The reduced sum was justified by an appraisal of the *gabinetto* carried out in 1827 by two "machinists" from the Pontificial University of Bologna, Don Luigi Poletti and Pietro Toldi, who certified that the instruments were in a bad state of repair and obsolescence and judged them to have become unusable. This pitiless diagnosis by the two technicians did not take into account the added value, both historical and symbolic, that the *gabinetto* had accrued as a result of having belonged to Laura Bassi. At any rate, after this appraisal there are no other traces of the *gabinetto* in the historical record. The Aldrovandi Marescotti family were forced to sell far more important properties, including the Camaldoli villa itself, and in the end they probably sold off the collection piecemeal.[7]

Fortunately, the inventory of the equipment remains. On the basis of this inventory, this chapter seeks to illustrate the specific structural and functional characteristics of the Veratti household's scientific *gabinetto* and the role that it played in the scientific and social life of eighteenth-century Bologna. Firstly, the inventory allows us to evaluate the size of the collection; while it cannot compete in size with those of eighteenth century public institutions or the collections assembled by wealthy *amateurs*,[8]

---

[4]Marta Cavazza (1995a), p. 737.

[5]Amongst the most recent and significant studies on Laura Bassi see Elio Melli (1988), Paula Findlen (1993, Gabriella Berti Logan (1994), pp. 785–912, Beate Ceranski (1996), M. Cavazza (2005), M. Cavazza (2009a), pp.115–128, M. Cavazza (2020).

[6]M. Cavazza (1995a), p. 734, note 39. For Dalle Donne, the first women to graduate in medicine at the University of Bologna (1800), see Gabriella Berti Logan (2003).

[7]M. Cavazza (1995a), pp. 737–739. For events concerning the Aldrovandi Marescotti family see Marina Calore (1994).

[8]A close example is Lord George Cowper's museum, purchased by the *Istituto delle Scienze* in Bologna at the end of the Eighteenth Century, see Giorgio Dragoni (1985).

it was still remarkable considering its owners' more limited buying power. Giovanni Fantuzzi's evaluation made the year of Laura Bassi's death (1778) was probably excessively cautious in defining the laboratory that she had formed "at her own expense" as a "not-unworthy collection of machines for physical experiments".[9] In reality, the importance and remarkable nature of this experimental apparatus depends not so much on the quantity and quality of the equipment (the richness of the materials, their size and the prestige of those who made them) as on its comprehensiveness and how well it served the uses it was destined for and the role it played in the scientific community of Bologna and Italy more broadly.

From the inventory, it can be deduced that the *gabinetto* was arranged according to didactic criteria. This is quite appropriate given that it was used by the owners for over forty years (mainly by Bassi for the first thirty years and then by her husband, after her death) to teach complete courses in experimental physics. The mechanisms and materials were grouped into eleven categories, each experimentally illustrating a specific discipline: "general properties of bodies" (extension, impenetrability, divisibility, figurability (*figurabilità*) and porousness), "attraction and repulsion", "magnetics", "gravity", "mechanics and motion", "hydrostatics and hydraulics", "atmospheric air and artificial air (*aria fittizia*)", "barometers, thermometers, hygrometers", "heat", "light", and "electricity".[10] We can glean an idea of the owners' interests from the richness of certain categories, in particular the large number of instruments designed to illustrate themes such as the physics and chemistry of gases (some of which purchased presumably after Bassi's death), light and optical phenomena, and electricity. The list also includes various kinds of eudiometers; mirrors, lenses and, above all, the prisms (British, "Icelandic", Venetian, made of glass or rock crystal) and apparatuses necessary to replicate Newton's experiments in optics; one of Franklin's multiple plate capacitors and two of Volta's electropheri; no less than nine microscopes and a considerable number of barometers, thermometers and hygrometers that probably were used for the meteorological measurements Verratti took both at home and at the Institute. Naturally, there were also the most spectacular instruments found in eighteenth-century laboratories—a pneumatic pump (with all of its accessories) for experiments on vacuums and "airs", and an electrical machine. Although this latter instrument was considered "ancient" (the Verrattis had bought it in 1746 and were only the second people in Bologna to own one, after the *Istituto*), it was the type "with two cylinders and two crystal globes" and could be connected to the first instrument to conduct experiments on the behaviour of "electrical fluid" in a vacuum: indeed, one of the two globes featured "a tap made in Holland to extract the air with the pneumatic machine".[11]

---

[9]Giovanni Fantuzzi (1778); then included in Giovanni Fantuzzi (1781–1790), pp. 384–391: the passage quoted is on p. 388, note. No. 9, of this latter edition.

[10]M. Cavazza (1995a), pp. 740–753.

[11]Marta Cavazza (1995a), p. 751. An older description of the machine can be found in Jean Antoine Nollet, *Journal du voyage de Piémont et d'Italie en 1749.* Soissons, Bibliothèque Municipale, ms. 150, p. 229: cfr. M. Cavazza (2009a), pp. 117–118 and note 13.

Most of the instruments and their related accessories were probably made in Bologna by specialised local technicians, mainly priests or members of religious orders, the same craftsmen also used by the *Istituto delle Scienze*.[12] The names of their designers or makers are specified for only a few of the machines. For instance, there is a "universal steelyard balance" and a pile driver made by Micheli, a "small Nollet mortar for the law on Projectiles", a "central forces machine described in the works of Nollet", a Musschenbroek tribometer, "a Boerhaave burner [...] for chemical operations with two retorts", an "objective lens" made by Montanari, an "English prism made up of three crown glass prisms for dolonian correction (*correzione doloniana)*", a Lyonnet microscope, a "Sauberf eudometer in graded crystal with Kunckel phosphorus", a "large barometer made in the manner of a clock with its thermometer in Brazilian wood manufactured by Malagida Seniore", several machines built by Volta including an electric pistol and an eudometer for inflammable air, two electrophori and a number of batteries. The inventory rarely indicates the place of origin; exceptions are the three microscopes in valuable materials imported from America by Francesco Veratti, Giuseppe's father, the prisms ground in England or Venice, the "electrical carillon made up of five metal bells made in Augsburg", or another instrument described as "a crystal pipe fitted with brass at the ends and that encloses a luminous conductor made in Florence". In at least one case, the information in the inventory is unquestionably inaccurate: the English compound prism mentioned above is described as having been "sent as a gift by Newton to Doctor Laura Bassi-Veratti",[13] which is impossible for obvious chronological reasons. Newton died in 1727, five years before Bassi earned the university degree that made her famous throughout Europe. Since we do not know when the imaginative note on the origins of the prism was written, it is difficult to establish whether it is part of the legend that grew up around Bassi as a public figure or, less innocently, an attempt to boost the historical and economic value of the *gabinetto*.[14]

Further information regarding the provenance of some instruments is available in the owners' correspondence. From Bassi's letters to Lazzaro Spallanzani we discover, for example, that the "Lyonnet microscope" mentioned above (a microscope equipped with a support for anatomical experiments) was made, thanks to the intervention by Spallanzani himself, by a skilled builder from Scandiano, Brother Fedele, and was meant to enable experiments on the reproduction of the heads of snails that Doctor Bassi had been carrying out for some time at the request of her

---

[12]Don Francesco Vittuari, a priest, the first "mechanic" hired by the *Istituto*, was the constructor of almost all the apparata used before 1744 (when the instruments ordered from Holland arrived) and again in the 1780s the secretary Sebastiano Canterzani pointed out the importance of the contribution made by members of the *Ordine dei Servi* (Servite Order) to the construction of up to date scientific instruments; cfr:. *De Bononiensi Scientiarum et Artium Istituti atque Accademia Commentarii* (from now on cited as *Commentarii*), Ex Typographia Laelii a Vulpe, Bononiae 1731–1791, volls. 7, I, (1731), p. 16 (for the information on Vittuari); VII, (1791), pp. 14–21.

[13]M. Cavazza (1995a), pp. 750–751.

[14]On the myth that grew up around the figure of Laura Bassi see Marta Cavazza (1995a), pp. 732–736.

former student.[15] We also learn that some of the prisms contained in the *gabinetto* were bought in England, commissioned by Bassi, from a Spanish expert on scientific instrumentation named Jose Hortega. It was through Hortega that Bassi also received a letter and special lens to reunite refracted rays, sent by Abbé Nollet.[16] Some years later, the famous physicist and instrument maker also sent her other devices he had created ("quelques bagatelles qui pourraient Vous amuser et dont Vous pourrez faire part aux amateurs de la physique qui fréquentent Votre *Museum*"), including a spectacular little machine that he had invented (similar to the "magical square with luminous strips" named in the inventory) that could be used to "conduire le feu electrique sur tel dessein qu'on voudra", thereby making different patterns of light appear on a square.[17]

Although they are not clearly identifiable on the inventory list, the correspondence between the couple in Bologna and Giambattista Beccaria shows that he sent two rock crystal prisms, included with a letter to Veratti, after describing at length in a previous letter to Bassi the "double refraction of rock crystal" and the relative law that he had discovered.[18] As for the Volta machines named in the catalogue, some (such as the inflammable air pistol) were certainly acquired during Doctor Bassi's lifetime and used by her personally, as we can see from her correspondence with the young inventor. Indeed, Volta had sent her booklets containing a description of the structure and operation of some devices that he had built, as a gift.[19] Others must have been bought instead by her husband or even by her son Paolo, such as the electric batteries which were invented after not only Bassi's death but also that of her husband.

[15]Spallanzani to Bassi, Modena, 30 April 1769; Bassi to Spallanzani, Bologna, 19 May 1771, in Pericle Di Pietro (1984), pp. 170–193. For Laura Bassi's role in Spallanzani education and on the relations between them I refer to Marta Cavazza (1999), pp. 185–202. On the modified Lyonnet microscope used by Spallanzani, cfr. Giulio Barsanti (2000); regarding Brother Fedele: Roberto Gandini (1972).

[16]Cfr. Antonio Garelli (1885), the letter from Hortega dated 21 April 1753 (pp. 70–80) and the letter from Nollet dated 30 March 1753.

[17]Antonio Garelli (1885), pp. 99–102, letter from Nollet to Bassi, undated. An extended version of this letter was published by Nollet in his *Lettres sur l'Electricité dans lesquelles on trouvera les principaux phénomènes qui ont été découverts depuis 1760*, printed in Paris in 1767 by L. Guérin et L. F. Délatour.

[18]Beccaria to Veratti, Turin 18 October 1766, in *Lettere di G.B. Beccaria a Laura Bassi e al marito,* Biblioteca Comunale dell'Archiginnasio di Bologna (from now on BCAB), Coll Autogr., VI, 1741–1754, n. 1747; Beccaria to Bassi, Turin, 18 May 1761, *ibid.*, n. 1747.

[19]M. Cavazza (1995a), p. 729.

6.2 Public Institutions and Private Teaching Sites in Eighteenth-Century Bologna

Although the facts that can be deduced from the inventory itself and the owners' correspondence are sparse, they do reveal one of the most significant aspects of an eighteenth-century scientific *gabinetto*, namely the fact that it was a place for not only research and teaching but also meeting up and exchanging views, in other words, a site of scientific and social relations. Indeed, in the main decades of the Enlightenment's spread, on a private level the Veratti household laboratory represented one of the main meeting places for the Bologna-based writers and scientists most interested in discoveries from abroad; it also constituted a landmark space for foreigners passing through the city. Although no precise documents have emerged so far about the structure of the Veratti living quarters, they presumably adjoined the parlour where "Madam Laura" received her visitors. Felice Fontana was one of the students who took advantage of the teachings of both husband and wife when staying in Bologna as a young man, and in later years he maintained an assiduous correspondence with them. According to Fontana, the couple's home was a place where one could meet "the most outstanding men of merit" in the city and it was "above all others one of the most courteous for persons of merit and foreigners".[20]

As an intellectual meeting place, the home of the Veratti couple originally stood out from the many other private houses hosting social encounters among Bolognese intellectuals and scientists by virtue of its history and characteristics. Firstly, it distinguished itself from the *Istituto* and attached *Accademia delle Scienze*. It is important to recall that, in eighteenth-century Bologna, these two institutions represented the centre around which all scholars interested in medical, natural historical and physical-mathematical disciplines gravitated. They constituted the authority that served to legitimate every scientific initiative, invention and theory.[21] They were supported in part by the generous and enlightened patronage of Benedict XIV, support which amongst other things allowed the institute to purchase a collection of cutting-edge equipment from Holland for the chamber of physics. In the middle of the century, the *Istituto* and *Accademia* showed considerable dynamism and enjoyed widespread prestige.[22] This probably contributed to spreading interest in science across urban milieu frequented by not only professional scientists but also men of letters (fascinated above all by the research into electrical phenomena that had become a recurrent *topos* in the poetry of the time), as well as aristocrats and members of the

---

[20]Letter from Fontana to Veratti, dated March 1766, BCAB. Coll. Autogr., XXIX, 8024–8031, n. 8031. For his part Spallanzani had written a year before that the "domus'" of Laura Bassi was considered by the most learned men "non aedes privatae mulietis sed doctrinae perfugium, ac sapientiae templum" (dedication to Laura Bassi of his dissertation *De lapidibus ab aqua resilientibus,* Montanari, Modena 1765, now in Pericle Di Pietro (1996), pp. 155–182: 157–158.

[21]Regarding the origin of the institutions in Bologna see Marta Cavazza (1990). For more detailed information on the activity of the *Istituto* and the *Accademia delle Scienze* the three volumes of *Anatomie accademiche,* respectively edited by Walter Tega (1986, 1987) and by Annarita Angelini (1993) are fundamental.

[22]A. Angelini (1993), pp. 207–238; M. Cavazza (1996), M. Cavazza (2008a).

bourgeoisie (including women) whose generalist educational backgrounds inevitably excluded them from the specialist debates taking place at Marsili's *Accademia*. The most relevant example is the *Accademia dei Vari*, founded by Count Filippo Carlo Ghisilieri. This academy was frequented by aristocrats and members of the middle classes, both religious and lay people, and was open to women as well. It frequently chose philosophical or scientific themes as the subject for discussion at its meetings. The founding members included several members of the *Accademia*; for instance, Laura Bassi and her husband were extremely active members.[23] Naturally, it was mainly the students of the philosophy department at the university of Bologna who displayed this interest in experimental physics. Paradoxically, however, the *Istituto* that should have compensated for the university's pedagogical shortcomings in this field from the moment of its inception was not able to offer the kind of extensive, effective education found at the time in other Italian universities, such as Padua.[24] Not even the availability of the cutting-edge Dutch instruments proudly displayed in the physics chambers ensured such courses, although the equipment did foster the significant development of research activity in this field. The reason for these lacks (also found in chemistry and other disciplines) lay in the *Istituto*'s *Costituzioni* laid down in 1711.[25] For the first time in Italy, this statute established experimental courses designed for University students, although carried out in a new, distinct institution. However, the rules prescribed that these courses should consist entirely of "exercises", in other words a "demonstration" of experiments but without the relative theoretical framework, as this latter was only supposed to be provided through the "lessons" of the university. Such demonstrations therefore could only be held on the day the University was closed, for two hours, from November to mid-August. The 1737 reform introduced modern methods and concepts in university programmes and improved, but did not resolve, the problems stemming from this irrational distinction between "exercises" and "lessons", a distinction which had been dictated by political considerations rather than scientific or educational ones.[26]

As was traditional in Bologna, the solution lay in home-based lessons that teachers gave to supplement the public lessons held at the *Archiginnasio*. Indeed, it was common practice in the city for university professors to give courses in their own homes focused on the subjects they officially taught or other similar subjects (most likely for a fee, although there is little documentation available). Iacopo Bartolomeo Beccari, Domenico Gusmano Galeazzi, Tommaso Laghi, and Luigi Galvani held private classes in logic, natural philosophy, chemistry, medicine, anatomy, and obstetrics. Giovanni Antonio Galli established a home-based school of obstetrics equipped with all the instruments and models necessary for practical demonstration. These

[23]Regarding the Vari, see Maria Grazia Bergamini (1996).

[24]In Padua in 1739 the first chair in experimental physics was established (entrusted to Giovanni Poleni) and the following year a well equipped public laboratory was inaugurated, for which cfr. Gian Antonio Salandin, Maria Pancino (1987). For an overall view of the situation in Italy and in Europe, cfr. John L. Heilbron (1982), in part. pp.157–259, and Marta Cavazza (1995b).

[25]*Le Costituzioni dell'Istituto delle Scienze eretto in Bologna sotto lì 12 Dicembre 1711*, printed notice in ASB, Assunteria d'Istituto, Diversarum, b 9, No. 5.

[26]On these matters I refer you to Marta Cavazza (1993).

classes were open not only to medical students but also to "*cerusici*" and "*mammone*", that is, practitioners of medicine and midwifery who had practical experience but no academic qualifications or training, the equivalent of barber-surgeons in Britain or "wise women" in France. Later, Anna Morandi taught a private anatomy school, first in her own home and then in the house of an aristocratic patron, Girolamo Ranuzzi, using wax anatomical models sculpted by her or her late husband Giovanni Manzolini as efficient teaching tools. Her lessons and creations aroused so much curiosity and admiration as to draw many people to her museum, not only students but also a number of eminent visitors. Some teachers also organized private academies and discussion circles which may no longer have been positioned in opposition to public teaching curriculum and methods as in the previous century, but probably met the need for spaces of debate that were more cozy, inclusive and informal than the *Accademia delle Scienze* sessions. Galeazzi organised in his own home "a private Academy that he named *Academy of the Non Experts*, which flourished with men of valour, Philosophers and Physicians of great prestige" while "every Friday of the year" Tommaso Laghi gathered around himself an academy "in which practical medical and anatomical matters were discussed, greatly to the advantage of the Physicians and young scholars that attended". Beccari's pupils continued to meet for decades in the Marchesini Academy, which took its name from being based in the home of the physician Ferdinando Marchesini.[27] Most of these initiatives were viewed by their organisers and the academic and civil authorities themselves as an integral part of the university teaching system, so much so that they constituted a "requirement" when requesting or obtaining salary increases. As we shall see, the laboratory and school in the Veratti home shared some of these characteristics while also having certain features that distinguished it from other such cases.

6.3 Laura Bassi's "Private Courses in Physical Experimentation"

The new lectureship in particular physics (*fisica particolare*) established by the 1737 university reform was assigned to Giuseppe Veratti, a medical graduate who was extremely interested in the new theories and scientific methods of the time. He immediately bought some equipment and began to offer supplementary lessons, obviously of an experimental nature, in his home. He went on offering these domestic lessons until he was appointed to the chair in medicine in 1750.[28] However, he had paved the way for the far more incisive initiatives launched by Laura Bassi, his

---

[27]The information about the home-based schools of Beccari, Galeazzi, Laghi, Galli, Morandi and Manzolini, and also about some of the academies organised by some of them and by Marchesini, are mainly derived from the relevant entries in G. Fantuzzi (1781–1790); the information about Galvani comes from Mario Medici (1845). There are important recent studies on the waxwork anatomical laboratory and the relative anatomy exercises by the Manzolini couple and then by Morandi alone. Amongst the most complete studies are: Miriam Focaccia (2008); and Rebecca Messbarger (2010).

[28]For Veratti's career, in the University and in the *Istituto*, see S. Mazzetti (1847), *ad vocem*.

colleague at the *Accademia* and, at least formally, at the University as well. In 1738 the two had wed. The couple went to live in an apartment in Via Barberia where they went on to spend their entire married life and it was undoubtedly here that they set up and equipped the laboratory.[29] In the years following her spectacular "graduation" in 1732, the fact that a young, unmarried woman frequented exclusively male locations such as the University and *Accademia*, or mixed ones such as the salons of the nobility, had provoked malevolent murmurs. However, even Bassi's decision to marry was criticised by those who believed that only status suitable for a woman who had dedicated her life to study was that of a nun.[30] Nonetheless, her position as a married woman allowed Bassi to participate actively in scientific and literary life. As early as 1738 she organised "an academy, or rather a literary conference" in her marital home "in which, two evenings a week, exercises were done of cases of philosophy, geometry, etc." and in 1739 she officially communicated this initiative to the members of the University senatorial *assunteria*, presenting it as part of her duties as a salaried lecturer. In the same document and for the same purposes, Bassi claims that "she continuously receives in her home foreigners, undertaking to answer the questions that she is asked and to hold with them literary discourse as it pleases them, often being required to hold formal disputes in her home for such occasions, with the participation of many Gentlemen and Men of Letters".[31] Charles de Brosses was invited to one of these "philosophical conferences" in 1739 and he gave a disillusioned account of it. The French traveller discussed with her the properties of magnets and "the singular attraction that electrical bodies have". According to De Brosses, these discussions were mainly aimed at highlighting the lady of the house's own doctrine and were therefore a sort of exhibition, not real philosophical disputes. Indeed, it was through these encounters that the "dottoressa" acted out the role of Bolognese Minerva, the living allegory of learned Bologna that the city authorities had assigned her in 1732. And yet, as Bassi's correspondence with the Rimini physician Giovanni Bianchi since before her marriage shows, the literary meetings taking place in the home of Laura Bassi's father also represented an opportunity for scholars (of all ages) interested in further developing their understanding of Newton's work to gather together.[32]

It would therefore seem that, in this first phase, the marriage simply allowed her to continue "to make a spectacle of herself"[33] at home but without scandal. It gave her an opportunity to display her philosophical and scientific education and dialectic ability, gifts so exceptional in a woman as to render her a marvellous and rare phenomenon, the object of the curiosity and admiration of visitors and a source of prestige for

[29]Cfr. the *Biografia* (pp. 11–39) foreword by A. Garelli to the *Lettere inedite* (1885), p. 32, note 3.

[30]Regarding the Bolognese *dottoressa*'s "battles" to assert herself as a scholar and as a university teacher, as well as the Studies cited in note 5, I refer to M. Cavazza (1997b).

[31]*Nota di "requisiti"* published in Elio Melli 1960, pp. 53–187: p. 87.

[32]Charles De Brosses (1986), Beate Ceranski (1994).

[33]This compelling expression is contained in a letter from Eustachio Manfredi dated 17 October 1736 published in a German numismatic journal, «Der Wöchentlichen Historischen Münz-Belustigung», in the 27 February 1737 issue, dedicated entirely to Laura Bassi, pp. 69–70. Regarding the phenomenon of the spectacular exhibition of female knowledge, typical of Eighteenth-Century Italy, see Marta Cavazza (2009b).

the City and its cultural institutions. On the other hand, considering she had been prevented from holding a regular public teaching course at the University, there was not much else she could do. Indeed, when appointing Bassi to the lectureship in Universal Philosophy in 1732, the senate had established *ratione sexus* that this new female lecturer would only give lectures at her superiors' command and on special, solemn occasions.[34] Bassi struggled at length to obtain permission to teach normally and continuously in the *Archiginnasio*, but she was obliged to fall back on lessons at home. In the "note of requirements" presented in 1748 to ask for a salary increase, she claims that she had "been ready to give lessons in the public University" and that she had taken active part in the annual anatomical function, but she also says that she had "taught part Geometry and part Philosophy for four years".[35] These were obviously lessons given at her home: all attempts to "re-establish the question of public lessons" having failed, feeling a responsibility "to the public" and not wanting to "remain completely useless",[36] lecturer Bassi implicitly accepted the social limits inherent to her gender and gave up on the idea of teaching regularly at the *Archiginnasio*.

What appeared to be a retreat was transformed into a great success. Starting in 1749, the enterprising doctor set up a real home-based school of experimental physics, transforming her simple, supplementary private lessons into a much broader initiative. This new project had ambitions of public utility as well, as it at least partly compensated for the great gap in Bolognese university teaching mentioned above, that is, the lack of an adequate course in experimental physics. This decision was the result of a lucid analysis of the situation, as Bassi explained a few years later.

> And yet Experimental Physics has become nowadays such a useful and necessary science, and we who were the first in Italy to receive it and cultivate it when the *Istituto* was opened, now must see with our blushes and also to the detriment of our University, that in all the others it is taught more than here with that method and breadth that is sought after for the advantage of the young, giving whole courses in several places; since such a thing could not be done at the *Istituto* because of the very few lessons that are given there according to the regulations established for it, I was moved to think of using my however-scarce ability to serve the public in these studies, believing that it would require a mediocre commitment and only for the young who wished to be initiated in this faculty. But now it has become so public that I cannot dispense myself without mentioning even the most learned persons who often love to see the experiments and especially the foreigners from north of the Alps who pass continuously through this city.[37]

Her courses were certainly more in-depth and comprehensive than the weekly courses at the *Istituto*. In 1755, writing once again to Flaminio Scarselli, her friend

---

[34]ASB, Senato, *Partiti*, ff. 49–50. Regarding Bassi's battles to be able to teach properly see Berti Logan (1994), in part. pp. 796–800.

[35]Elio Melli (1960), p. 128.

[36]Elio Melli (1960), pp. 103–104: letter to F. Scarselli, Bologna, 21 April 1745.

[37]Elio Melli (1960), pp. 150–151: letter to R Scarselli, Bologna, 16 July 1755.

but also and more importantly the Secretary of the Embassy for the Bolognese Administration to Pope Benedict XIV, Bassi offers a very precise description.

As for my studies, it is now six years since I began to give private courses in physical experiments in my home, giving eight months of daily lessons accompanied by experiments; and having had all the necessary instruments made at my own expense, as well as those prepared by my husband when he was lecturer of Philosophy. The thing has grown so much that, instead of the young, more experienced persons and very often diligent foreigners come to these courses, so that I see myself needing to think about increasing the equipment and especially the more complex machines.

The *dottoressa* asked her correspondent for advice on how to reach "Our Lord" the Pope, to convince him to lend.

[...] A hand in this business, to which I have already committed more than 400 *scudi* and I still go on committing myself to more, but I see that I cannot accomplish what would be necessary without compounding this for my family, and I would still wish to do everything possible to prepare all the materials, so that those that continue to engage in this profession after me will at least be free of all the embarrassments that these apparatus bring with them.[38]

Scarselli, who was thoroughly familiar with the Pope's beliefs about the role public authorities should play in promoting science, did not present the *dottoressa*'s request. He anticipated that the Pope would undoubtedly reply that, "having spent, and still spending, so much to provide the *Istituto* with machines and other equipment, he should not think to provide more to private Homes".[39]

As these documents suggest, the fact that Laura Bassi's laboratory was necessary for her home-based physics school meant that it was both an economic investment, made in agreement with her husband, and the only opportunity for her as a woman to carry out the "duty to the public" that came with holding a university chair, a chair that was "extraordinary" but salaried. Her courses were not "coaching" or supplementary lessons for students; they were the only way for her to teach a complete course and the only means for students to gain familiarity with theoretical concepts and experimental methodologies that had by then become essential.

As a result of this importance, alongside her teaching skills, ability as an experimenter and up-to-date theoretical competence, her lessons met with great success and attracted many "pupils" from both Italy and abroad, in particular Greece, Germany and Poland.[40] Her students included the great naturalist Lazzaro Spallanzani, the mathematician Gian Francesco Malfatti, the future director of the Florence Natural History Museum Felice Fontana, the economist Gian Maria Ortes, and the botanist Casimiro Gomez Ortega (who would become professor and director of the Botanical Gardens in Madrid).[41] Moreover, from 1766 until her death she was appointed to

[38]Elio Melli (1960), pp. 148–149: letter to F. Scarselli, Bologna, 14 June 1755 (400 *scudi* corresponded to 2000 *lire*).

[39]Antonio Garelli (1885), letter from Scarselli to Bassi, Roma, 18 October 1755, pp. 122–124.

[40]G. Fantuzzi (1781–1794), I, p. 388, note 9.

[41]M. Cavazza (1999); Giuseppe Venturoli (1811); Antonio Garelli (1885), p. 80 (letter from Jose Hortega, Casimiro's uncle, from Madrid, 21 April 1753).

teach the boarders of the Collegio Pontificio Montalto. Spallanzani had been one of her first students in 1749 and even after her death he claimed that the original source of all his knowledge was the "wise teachings of his revered teacher".[42] It was Spallanzani who provides us with the earliest description of her lessons, attended, he says, by a large number of young men from neighbouring and more distant regions alike, and made fascinating by her remarkable brilliance as an experimenter and by the elegance of her speech.[43] The liveliest reconstruction of her teaching method is provided by a former Montalto college student in the eulogy delivered June 1778, a few months after her death.

You would have seen her surrounded by a numerous circle of pupils hanging on to her every word, giving the theory before the established experiments, the most exact, the richest in physical erudition, the most judicious, and giving it with an ordinary clarity, and at the same time with an elegance and richness of language which left listeners in doubt as to whether she was speaking spontaneously, as she often did, or whether she had meditated at length and written out what she pronounced. Then from the theory she passed to the experiments, and here she would operate everything with the most scrupulous accuracy, note the slightest differences, reveal those circumstances that most demonstrate the truth of the Phenomenon, form over it reasoning worthy of that great mind, [and] draw from it the most correct consequences.[44]

Bassi's lessons in experimental physics therefore differed from the *Istituto* "exercises" not only in terms of their daily frequency, completeness and systematicity, but also because she did not merely demonstrate unadorned experiments; she also granted great importance to outlining the theoretical framework that such experiments confirmed or disproved. In other words, she wholly rejected the absurd separation of theory and experiment that the university and *Istituto* teachers had to abide by. For its part, the senate fully recognised the public utility of Laura Bassi's private courses, and in 1759 it raised her lecturer's salary to one thousand *Lire* (rather a large sum, comparatively) "with the requirement of continuing to teach experimental physics at home, in the way that she teaches it and has taught it for the past ten years".[45] This raise was also understood as compensation for the often considerable expenses she incurred in purchasing and maintaining the laboratory instruments, as she did not fail to point out on many occasions.

---

[42]Letter from Spallanzani to Veratti, dated Pavia, 30 April 1782, in Pericle di Pietro (1984), XI, p. 71; Antonio Garelli (1885), pp. 218–219.

[43]Dedication *Laurae Bassiae, spectassimi nominis muliebri* [...] to Spallanzani's first work, *De lapidibus ab aqua resilientibus,* in *Dissertazioni due dell'abate Spallanzani*, Eredi B. Soliani, Modena 1765, pp. unnumbered (republished, in Pericle Di Pietro (1996), pp. 155–182).

[44]Ignazio Odoardi, *Elogio,* in *Pubblica Accademia di lettere avutasi nel Collegio Montalto dagli alunni del medesimo la sera delli 5 giugno 1778 in lode della defunta loro precettrice Laura Bassi* (BCAB, ms. B, 2727, cc. unnumbered).

[45]ASB, Senato, Vacchettoni, reg. 71, c. 95, *Aumento alla lettrice Laura Bassi*, 14 December 1759. Veratti also applied to the Senate and obtained reimbursement for the expenses sustained for experiments "of the kind that are carried out at home by Sig. Veratti for the Medical use of the Electric Machine": ASB, Assunteria. di Studio, Atti, 4, f. 1284. I thank Paola Bertucci for this reference.

6.4 A Family Laboratory but Open to the World

In 1776, Laura Bassi was appointed professor of Experimental Physics at the *Istituto*, with Veratti as substitute. This appointment was finally without any limitations *ratione sexus* and it marked the final victory in her struggle to teach in public with the same rights and duties as her male colleagues. Nevertheless, the reform to make the *Istituto*'s courses more complete and effective still had not been passed, and so her private teaching was still necessary. It was so important, in fact, that when she died two years later, her husband not only replaced her in the institutional role of professor of experimental physics but decided to continue the home-based school as well. Veratti updated some of the laboratory instrumentation to bring it in line with new frontiers in the study of electricity and the "airs" that he and his wife had devoted themselves to for years, as indicated by the reports they both presented to the *Accademia.* From 1761 to 1778, Bassi presented her academic colleagues with no fewer than five reports about electrical phenomena and two about gases, alongside her more usual investigations into rational mechanics, fluid dynamics and optics.[46] Unfortunately, the only of these to have survived are the two published in the *Commentarii* of the *Accademia (De problemate quodam hydrometrico,* vol. IV, 1757, pp. 61–73; *De problemate quodam meccanico, ibid.*, pp. 74–79); and two manuscripts, one on the air contained in liquids, presented at the *Accademia* in 1747 and 1748, currently stored at the *Accademia delle Scienze di Bologna* archive (Antica Accademia, Titolo IV, Sez. I, *Manoscritti delle Memorie dell'Accademia 1746–1753*) and now published in the appendix to Beate Ceranski's study. The content of this treatise, titled *De immixto fluidis aere,* is summed up by the secretary Canterzani in volume VII (1791) of the *Commentarii* (pp. 44–47). It was Ceranski herself who carried out a more detailed analysis of the reports on mechanics, the elasticity of air and air bubbles in liquids.[47] In the same period and after the death of his wife, Veratti engaged issues in both fields, moved mainly by medical interests and employing an original approach aimed at understanding the effects of electricity on organisms (see, for example, *Sopra l'accelerazione del polso e la materia elettrica,* 1765; *Sopra l'elettricità riguardo agli animali,* 1770) and the action of gases on biological liquids (*Sull'aria fissa,* 1776; *Sull'aria nitrosa,* 1777; *Sull'aria fissa contenuta nel latte,* 1780; *Sopra il color florido che acquista ii sangue essendo esposto all'aria,* 1781; *Su varie arie mefitiche,* 1782; *Sopra l'acido del carbone,* 1783). Only the essay on the effects of electricity in animals (*De animalibus electrico ictu percussis)* is referred to in the *Commentarii* (vol. VII, 1791, pp. 41–44), whereas we know the dates and titles of the others thanks to *Academia* records.[48] The couple's research into "fixed air" or carbon dioxide, a subject that aroused great interest in Europe, was far from common in Bologna. As Sebastiano Canterzani wrote in 1776, "[...] no one thinks of fixed air other than Signor Veratti and Signora Laura, who have carried out experiments on it"

---

[46]See the list of the titles of the 32 essays presented by Bassi to the Accademia in E. Melli (1988), p. 79. Cfr. also A. Angelini (1993), Appendix I. *Il diario scientifico*, pp. 313–448, *passim.*

[47]Beate Ceranski (1996), pp. 138–161, pp. 260–270 (Appendix).

[48]Cfr. A. Angelini (1993), *Il diario scientifico*, cit. in note 46, *passim.*

even though they considered it still obscure as a phenomenon, with "theory" of which had yet to be developed.[49] The list of instruments in the inventory's "Atmospheric air and artificial air" section contains no less than 37 items, testimony to the growing importance of this field of study for Bassi and Veratti.[50]

The experiments planned for these and other investigations were carried out by both of them in their home laboratory. Beginning in the 1740s when they first set it up with a considerable economic outlay and untiring passion, this laboratory had served their respective research: Veratti's on medical electricity, the analogy between electrical and magnetic phenomena, and atmospheric electricity published in his 1748 book,[51] and Bassi's on the general validity of Boyle's and Mariotte's law regarding the ratio between volume and pressure in gases, the dynamics of fluids, and the phenomenon of the bubbles that form in liquids in the absence of pressure.[52] Regardless of the differences in their disciplinary fields, both of their theoretical leanings were characterised by a steady adherence to epistemology and the Newtonian method. Such facts lead us to imagine that there must have been a constant and fruitful exchange of ideas between the two scholars, undoubtedly encouraged by their sharing daily life, the space and the instruments they used.[53] The availability of these instruments, and in particular the electrical machine they bought in 1746, also facilitated their establishing collaborative and idea-exchanging relationships with other scholars. Indeed, these relationships were so lively that the scientific *gabinetto* in the Veratti household became a theatre for public demonstrations of experiments, either new or already known, controversial physical or medical discussions, the centralisation and dissemination of news and books, and encounters both among foreign scholars passing through Bologna and between established scientists and young, novice scholars. Above all, it became a point of reference for Bologna-based thinkers who had not only subscribed to Newtonian physics and cosmology but were also more open to the new Newtonian developments in sectors such as physiology and electrical physics. In the mid '50s, the Bolognese scientific community was divided between supporters of Haller's new physiological theories, concentrated on distinguishing between sensitivity and irritability and denying the role of animal spirits as mediators between nerves and muscles, and adherents to the rigid mechanical tradition of Malpighi. The Veratti couple openly sided with the former of these positions and made their laboratory available to Caldani and Fontana, fierce young advocates of Haller, to conduct their experiments. With figures such as Francesco Algarotti and

[49]Canterzani to Bartolomeo Mozzi, Bologna, 7 December 1776, Biblioteca Universitaria di Bologna, ms. 2096, b. 6.

[50]Cfr. M. Cavazza (1995a), pp. 745–747.

[51]Giuseppe Veratti (1748, 1752). For an overall view of the two scholars' interest in electrical phenomena I refer to G. Berti Logan (1994), pp. 808–810; B. Ceranski (1996), pp. 162–169. M. Cavazza (2009a), pp. 115–128; regarding the controversy over the medical use of electricity and Veratti's role see Paola Bertucci (2005), pp. 64–100 and 123–171.

[52]As well as the experiments described in the dissertations published in the *Commentarii* and in the manuscript essays, there are those reported by the Secretary of the *Accademia* in *Commentarii*, t. II, part I, 1745, pp. 347–353, with the title *De aeris compressione*.

[53]B. Ceranski (1996), p. 169.

the surgeon Pier Paolo Molinelli in attendance, these experiments focused in particular on the electrical stimulation of "irritable" and "sensitive" parts, procedures that absolutely required an electrical machine.[54] Bassi and Veratti were not only amongst the first scholars in Italy to study electrical phenomena, they were also amongst the first and most faithful supporters of the theory of the single electrical fluid with opposed polarities first proposed by Franklin and then systemised by the Turinese professor Giambattista Beccaria in the framework of Newtonian physics.[55] The two Bolognese scholars did not merely exchange letters with Beccaria. Indeed when he came to Bologna in October 1755 to illustrate his theories, they had the opportunity to work directly with him in the physics chamber of the *Istituto*, participating in replicating experiments that had already been published in *Elettricismo naturale e artificiale* (1753).[56] During the Turinese physicist's stay in the city, he frequented the Veratti home and laboratory assiduously, repeating experiments both for and with his hosts. Some of these were already known but they also invented new ones together, innovations which he then described in letters to Bologna *Istituto* president Iacopo Bartolomeo Beccari, making up the work *Elettricismo naturale*. Amongst other things Beccaria described an experiment with the aim of confirming the "universal diffusion of electrical vapour" proposed by the "valiant Lady Laura, who, in truth, never dislikes good reasoning but is always pleased by experiments".[57] He continued to correspond with his two Bolognese colleagues and Bassi in particular. Indeed, he went on to ask her to act as his official intermediary with the *Accademia* and proclaimed his gratitude to her for the "expressions that you use, with all that pass that way, both with foreigners and with our countrymen above all, speaking of my work and of my theory".[58] On at least two occasions Bassi personally acted as a spokesperson in presenting Beccaria's discoveries and theories to the *Accademia*: when she communicated his explanation of the double refraction of rock crystal and when she illustrated his idea of *electricitatis vindex* (avenging/defending electricity) that is, the corrected version of Franklin's theory that Beccaria had introduced to rebut Robert Symmer's criticisms.[59] And it was certainly due to Bassi as well as Veratti

[54]Marc'Antonio Leopoldo Caldani, *Sull'insensitività ed irritabilità di alcune parti degli animali. Lettera scritta al chiarissimo e celebratissimo signore Alberto Haller*, Bologna 1757, pp. 269–336, in part. p. 325, where the author describes an experiment on the electrical stimulation of the fibres of the heart in which Bassi actively participated and in which he thanked Veratti for having suggested to him the "new method" followed in these electrophysiological experiments. An interesting interpretation of the episode has been proposed in Paula Findlen (2003). On the debate between Hallerians and advocates of the Malpighian iatromechanical tradition, I refer to Marta Cavazza (1997a, 2008b).

[55]For an extremely useful text concerning the studies on electricity carried out in Italy, see Antonio Pace (1958).

[56]Giambattista Beccaria (1753). Regarding Beccaria's stay in Bologna and his influence on researchers into electricity at the Accademia, see N. Urbinati, *Physica,* in Walter Tega (1987), pp. 123–154.

[57]Giambattista Beccaria (1758).

[58]Beccaria to Bassi, Turin, 26 [...] 1759, BCAB, Coll. Autogr., VI, 1741.

[59]In a letter to Veratti dated 10 October 1763 Beccaria sent his thanks to Bassi "for the commemoration that she chose to make of my law of the refraction of rock crystal " (*ibid.*, 1745), referring

that the *Istituto* in Bologna never came under attack by Symmer's followers with their doubts as to the unicity of electrical fluid. This adherence to Beccaria's current of thought was so solid, in fact, that the 1780 official *Istituto* guide stressed that the instrumental apparatus of the electricity chamber featured "the items necessary to explain Franklin's and Beccaria's system".[60] Felice Fontana was another figure who had harboured deep esteem for the couple since staying in Bologna as a young man, when they not only gave him the opportunity to use their *gabinetto* to carry out experiments on Haller's irritability but also supported him and advised him with his innovative research on the iris[61]; in the end, however, not even Fontana was able to convince them. He wrote Laura Bassi from Florence outlining his doubts about Franklin's system but, despite some momentary uncertainty, she did not abandon her convictions. Nor did her husband, and indeed in 1778–79 and 1779–80 he went on to dedicate the experimental lessons held at the *Istituto* "to Beccaria's and Franklin's system".[62]

There is no doubt that it was Doctor Bassi who constituted the hub of the network of relationships branching out from the home in Via Barberia and the main motor of the group that frequented it. Veratti remained somewhat in the shadows despite the esteem he enjoyed in the Bolognese scientific community and his important role in breaking new ground in research that later led Luigi Galvani, another figure who frequented the Bassi-Veratti laboratory assiduously, to discover animal electricity.[63] Veratti had his own network of correspondents, but it was often his wife who passed on his messages and informed their common correspondents about his research or the research they carried out together. Undoubtedly Bassi's leading role stemmed in part from her visibility (indeed, overexposure) as a result of the Bolognese authorities' making much of her scholarship, extraordinary for a woman, in their propaganda since 1732. Some of the personages coming from other Italian cities, Europe and even America to exchange ideas with her or observe one of her experiments were surely motivated by curiosity at such an exceptional phenomenon. However, her visibility can also be explained by certain gifts that Bassi possessed in abundance, talents that made her perfect for presenting the "sociable" vision of science typical of the "enlightened century" in which she was proud to live.[64] Her correspondence, both

---

probably to the essay *Sopra il vetro islandico*, now lost, which she presented to the Accademia on 29 April 1762 (cfr. A. Angelini (1993), vol III, p. 352). Bassi's lost essay with the title *Sopra l'elettricità vindice* presented in the *Accademia* on 7 June 1771 (*ibid*., p. 363) must certainly have been a speech in favour of the new version of Franklin's theory proposed by Beccaria.

[60] G. Angelelli (1780). On the controversy between the supporters of Symmer and those of Franklin and Beccaria, and on the first studies on electricity in Italy, see J. L. Heilbron (1982), pp. 320–321 and 343–348.

[61] Regarding this see Fontana's letter to Veratti dated 25 March 1759 from Florence, in Antonio Garelli (1885), pp. 211–221.

[62] For Bassi, see Fontana's letters dated 10 June 1768 and 9 May 1775 from Florence (BCAB, Coll. Autogr. XXIX, 8024 e 8028). For Veratti «Diario Bolognese Ecclesiastico e Civile», years 1779 and 1780.

[63] For the influence of Veratti's research on Galvani's development see Marco Bresadola (2011).

[64] Bassi to Spallanzani, Bologna, 9 April 1768, in Pericle Di Pietro (1984), X. p. 223.

published and unpublished, documents the variety of relationships she maintained with other researchers, patrons and intermediaries, and young men who turned to her as a go-between to develop contacts with the Bolognese *Accademia*. Many arrived in Bologna armed with letters of presentation addressed to her and signed by other correspondents of hers. Some sent books, not only for her personal collection but also so that she would make them available to her colleagues at the *Accademia*, particularly the secretary and president.[65] With some of her correspondents, including Nollet, Fontana, Spallanzani and Beccaria, she also engaged in an intense exchange of instruments and/or information about such instruments. With Spallanzani, who had been one of the first to frequent her school, she developed an interesting collaboration at a distance, agreeing to carry out experiments under his direction in order to corroborate the experiments he himself had carried out on the reproduction of the heads of snails.[66] In his letters, Fontana who had been hired as director of the public museum in Florence outlined the new instruments the museum had purchased. He even sent her an example of a "little machine" built in Florence under the guidance of its inventor, the Dutchman Jan Ingenhousz (whom Bassi herself had introduced to Fontana) so that she could see it and have it reproduced.[67]

From at least the 1760 s onwards, Laura Bassi's fame and reputation in not only Bologna but throughout Italy no longer derived from the exceptional fact of her being a *femme savante* and having taken on male roles; rather, it reflected the admiration she had earned within the scientific community. Young men interested in electrical and pneumatic phenomena in particular must have seen her as an authority (from 1776 she was also invested with formal authority) from whom to seek advice and above all support in their efforts to gain standing and acceptance in the community of Italian physicists, a community of which the Bolognese *Accademia* was one of the most important expressions. For instance, we can identify three cases of scientists who were certainly not followers of the Franklin "party" by chance. One was Marsilio Landriani, who asked Bassi for her opinion regarding a new type of portable barometer that he had invented.[68] Then there was Giuseppe Campi, who sent her the collection of Franklin's works, the first translated into Italian, that he had edited in 1774.[69] The third and most famous of Laura Bassi's correspondents, Alessandro Volta, also divided his attention between electricity and the chemistry of "airs".[70] On Spallanzani's advice, when Volta was still a young and little-known inventor in 1771 he sent her a short work containing the description of a series of new electrical

[65] Cfr. Antonio Garelli (1885) *passim*; Elio Melli (1960) *passim*. For a more extensive examination of Bassi's role in the scientific community of her time I refer to M. Cavazza (2005).

[66] M. Cavazza (1999), pp. 195–197.

[67] Fontana to Bassi, letters dated 8 February 1771 and 30 April 1775, from Bologna (BCAB, Coll. Autogr., XXIX, 8027 e 8029).

[68] Landriani to Bassi, Milan, 4 July 1777 (BCAB, Coll. Autogr., XXXVII, 10054).

[69] Campi to Bassi, letter dated 8 August 1774, from Milan (BCAB, Coll. Autogr., XIII, 3868).

[70] Regarding Volta's contribution to pneumatic chemistry see the articles by Ferdinando Abbri, Bernardette Bensaude-Vincent, Raffaella Seligardi, Marco Beretta, Fredric. L Holmes in Fabio Bevilacqua and Lucio Fregonese (2000). Regarding Volta's intellectual biography see Giuliano Pancaldi (2003).

experiments. In 1776 (via Campi) he then made a point of showing her his first writings on the inflammable air of marshes, followed by the entire work on this subject a year later. In 1777 he sent her his leaflet describing his inflammable air pistol and gave her advance notice of a new invention, the inflammable air lamp, the description of which he intended to dedicate to her: "a fair ornament of natural sciences, and light and glory of her sex in our Italy".[71] In these letters from Volta and the only one of Bassi's replies to have survived, it is striking how much excitement the now elderly *dottoressa* expresses about the inventions of her young correspondent.[72] Such enthusiasm was certainly shared by Veratti as well. Even after his wife's death, he continued to acquire new Voltian instruments for the laboratory that they had set up together in their home. The house itself had evolved from a space of domestic affection to a workplace and site for scientific exchange recognised by the *Europe savante* without ever losing the informal cordiality typical of a family home, no doubt thanks to the authoritative and amiable presence of "Signora Laura".

6.5 Final Conclusions

The scientific *gabinetto* the scientific couple created in the eighteenth century represented a forerunner of many others to come in later centuries. At the time, it was both unique and perfectly reflective of the new forms of social relations that were giving rise to a historical shift, taking natural philosophy out the universities and academies and placing it instead in the "circles" and "tolette delle dame" of the time, just as their friend Algarotti had wished in the 1730s.[73] The discussions that scholars, students, instrument makers and curious onlookers engaged in as they populated the home and laboratory in Via Barberia were certainly poles apart from the amateur, glib conversations about issues of scientific interest such as electricity, magnetism and vacuum taking place in salons. Nevertheless, they were both a manifestation of the same social and cultural climate.[74] In reality, the innovative character of this example lies not one single aspect but rather in the multiplicity of functions performed by a private laboratory like that of the Veratti couple. It was a structure dedicated to both experimental research and teaching. It was a point of reference for innovators, both young and less so, with their gaze fixed on Europe. It was also a space for discussing, empirically verifying and disseminating theories and experimental techniques that

[71] Volta to Bassi, letters from Como dated 15 July 1771 and 15 June 1777, in Antonio Garelli (1885), pp. 157–158 and 158–159; Campi to Bassi, letter from Milan dated 28 November 1776 (BCAB, Coll. Autogr., XIII, 3869).

[72] Bassi to Volta, Bologna, 10 September 1777, in Francesco Massardi (1949).

[73] On couples in science see the essays collected in Helena M. Pycior, Nancy G. Slack, Pnina G. Abir-Am (1996); Francesco Algarotti (1737); on the transformation of places for experimentation in the Age of Enlightenment see the contributions to the volume edited by Bernardette Bensaude-Vincent and Christine Blondel (2008).

[74] For these questions see P. Bertucci (2005), pp. 172–215.

stood at the centre of an international network fed by both Bassi's intense correspondence and, to a lesser extent, Veratti's letters as well; at the same time, the network was also fuelled by the accounts provided by Grand Tour travellers as they arrived and departed. And finally, last but certainly not least, it served as an organisational hub for academic politics, from promoting new publications to manoeuvring for appointments to lectureships and academic posts in Bologna and elsewhere, for the couple themselves and for friends and pupils.[75] Some of these functions were a private version of the tasks that Luigi Ferdinando Marsili had assigned to the *Istituto* and *Accademia delle Scienze*. Others, like the formation of homogeneous groups of researchers, had deep roots in the history of the university but may also have represented a precursor to developments in the nineteenth and twentieth centuries.

None of the private schools and academies active at that time in Bologna were characterised by such complexity or degree of vitality. The transformations underway throughout Europe regarding practices for transmitting and disseminating Newtonian experimental science, as well as developing relations between scientists and the public, were certainly favourable factors.[76] Another factor not to be underestimated is the specific social and cultural dynamics that came to permeate Bologna during the so-called Lambertinian age, that is to say the reign of Pope Benedict XIV. In both of these macrophenomena, the new role assumed by women was both significant and highly controversial. There is no question that the main factor enabling a project such as the Veratti family's laboratory was the unprecedented step of admitting a woman to social and professional roles that previously had been exclusively male. The second condition to have created this opportunity was Giuseppe Veratti's confident participation in this endeavour of creating a family business the economic implications of which were at least as important as its cultural ones. Indeed, his stance implicitly paved the way for innovative new gender relations. Nonetheless, the final ingredient in the success of this project was undoubtedly Laura Bassi's intellectual authority, personal charisma and lucid determination.

Figures 6.1, 6.2, 6.3, 6.4, 6.5, 6.6, 6.7, and 6.8 show documents relative to the Bassi-Veratti couple and some laboratory instruments of the time.

[75] Examples of these academic "manoeuvres" can be found in the correspondence in Antonio Garelli (1885), *passim*, and in Elio Melli (1960), *passim*.

[76] Amongst the many articles on the subject I restrict myself to citing Moira R. Rogers (2003).

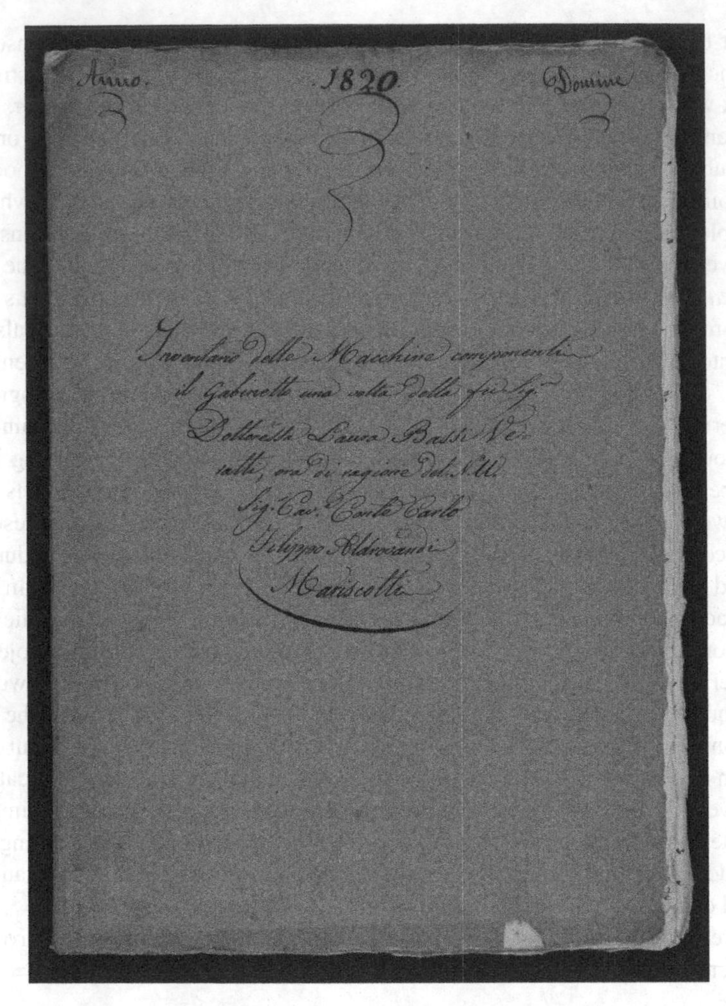

Fig. 6.1 Frontespice of the manuscript booklet listing the components of Laura Bassi-Veratti's Laboratory, Archivio Aldrovandi Marescotti, b. 430, c. 23, Archivio di Stato di Bologna

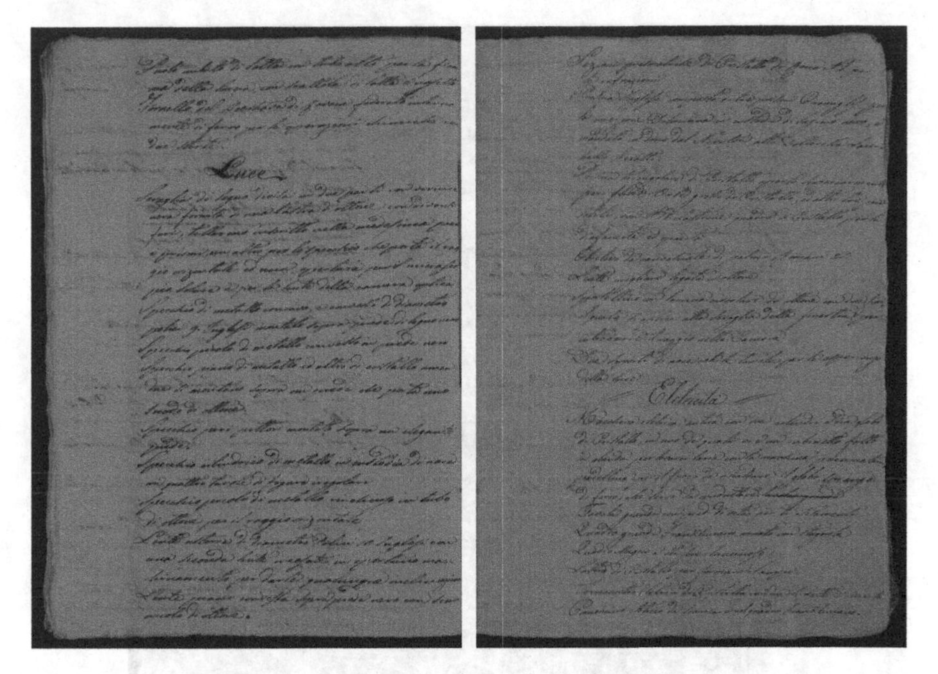

Fig. 6.2 Inventory pages corresponding, respectively, to the entries "Light" and "Electricity", Archivio Aldrovandi Marescotti, b. 430, cc. 14, 19, Archivio di Stato di Bologna

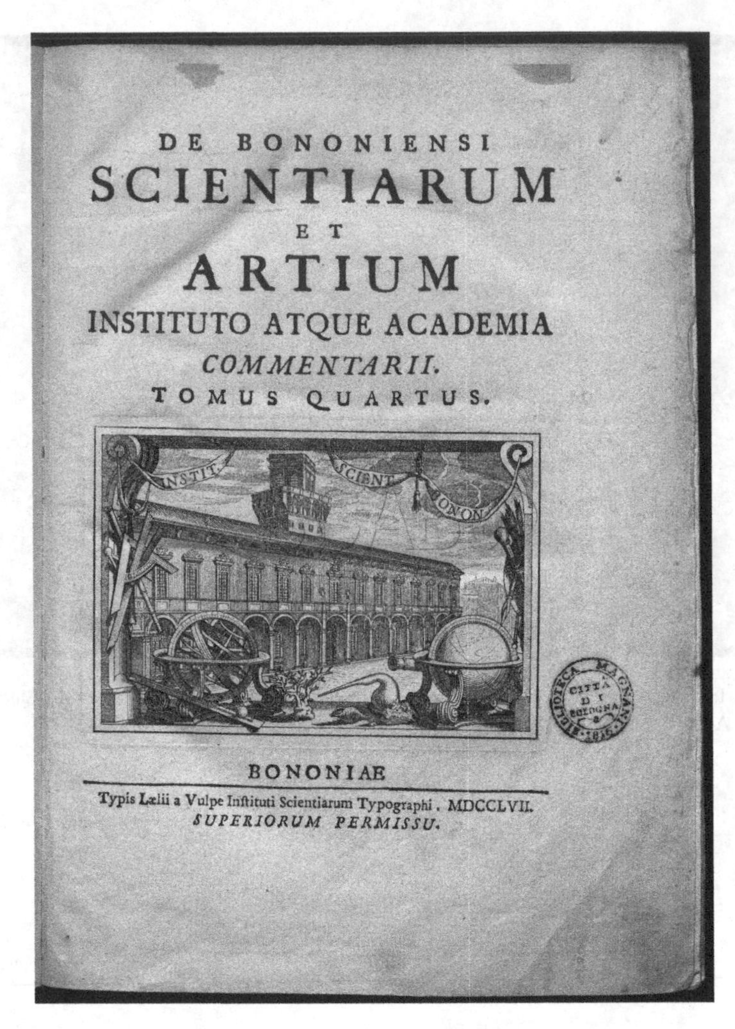

DE BONONIENSI
SCIENTIARUM
ET
ARTIUM
INSTITUTO ATQUE ACADEMIA
COMMENTARII.
TOMUS QUARTUS.

BONONIAE

Typis Lælii a Vulpe Inſtituti Scientiarum Typographi. MDCCLVII.
SUPERIORUM PERMISSU.

Fig. 6.3 Frontespice of *De Bononiensi Scientiarum et Artium Instituto atque Academia Commentarii*, Bononiae, L. a Vulpe, 1731–1791, tome IV (1757), Bologna, Biblioteca Comunale dell'Archiginnasio

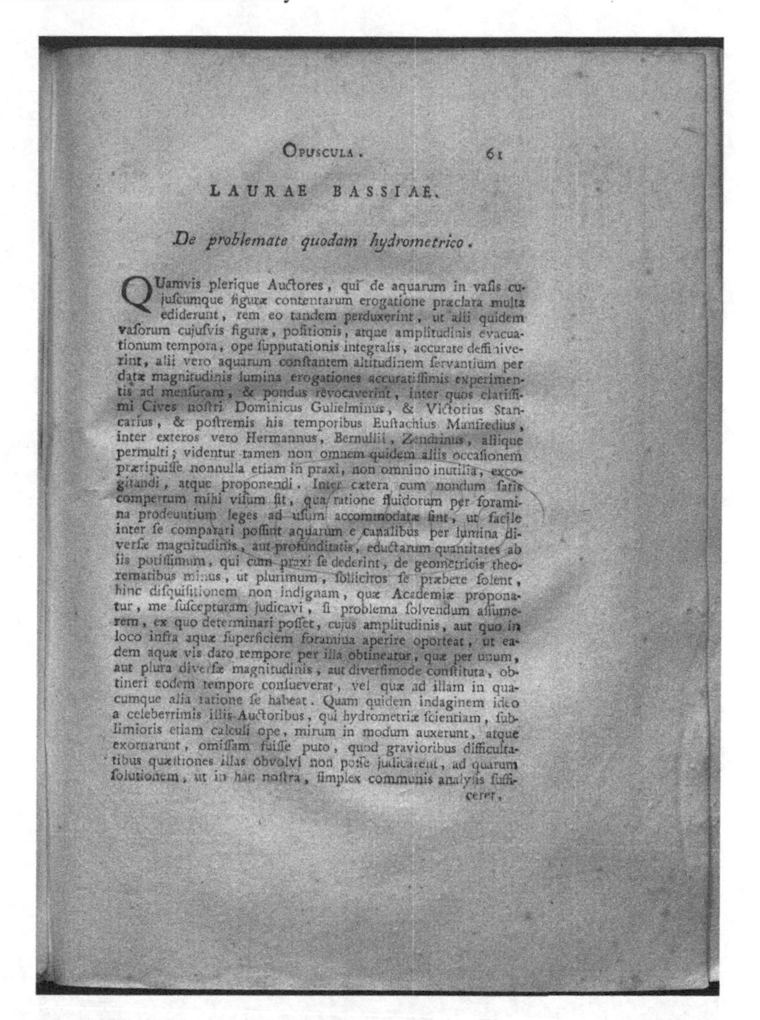

Opuscula. 61

LAURAE BASSIAE.

De problemate quodam hydrometrico.

Quamvis plerique Auctores, qui de aquarum in vaſis cujuſcumque figuræ contentarum erogatione præclara multa ediderunt, rem eo tandem perduxerint, ut alii quidem vaſorum cujuſvis figuræ, poſitionis, atque amplitudinis evacuationum tempora, ope ſupputationis integralis, accurate deffiniverint, alii vero aquarum conſtantem altitudinem ſervantium per datæ magnitudinis lumina erogationes accuratiſſimis experimentis ad menſuram, & pondus revocaverint, inter quos clariſſimi Cives noſtri Dominicus Gulielminus, & Victorius Stancarius, & poſtremis his temporibus Euſtachius Manfredius, inter exteros vero Hermannus, Bernullii, Zendrinus, aliique permulti; videntur tamen non omnem quidem aliis occaſionem præripuiſſe nonnulla etiam in praxi, non omnino inutilia, excogitandi, atque proponendi. Inter cætera cum nondum ſatis compertum mihi viſum ſit, qua ratione fluidorum per foramina prodeuntium leges ad uſum accommodatæ ſint, ut facile inter ſe comparari poſſint aquarum e canalibus per lumina diverſæ magnitudinis, aut profunditatis, eductarum quantitates ab iis potiſſimum, qui cum praxi ſe dederint, de geometricis theorematibus minus, ut plurimum, ſolliciros ſe præbere ſolent, hinc diſquiſitionem non indignam, quæ Academiæ proponatur, me ſuſcepturam judicavi, ſi problema ſolvendum aſſumerem, ex quo determinari poſſet, cujus amplitudinis, aut quo in loco infra aquæ ſuperficiem foramina aperire oporteat, ut eadem aquæ vis dato tempore per illa obtineatur, quæ per unum, aut plura diverſæ magnitudinis, aut diverſimode conſtituta, obtineri eodem tempore conſueverat, vel quæ ad illam in quacumque alia ratione ſe habeat. Quam quidem indaginem ideo a celeberrimis illis Auctoribus, qui hydrometriæ ſcientiam, ſublimioris etiam calculi ope, mirum in modum auxerunt, atque exornarunt, omiſſam fuiſſe puto, quod gravioribus difficultatibus quæſtiones illas obvolvi non poſſe judicarent, ad quarum ſolutionem, ut in hac noſtra, ſimplex communis analyſis ſufficeret,

Fig. 6.4 First page of the paper, *De problemate quodam idrometrico*, by Laura Bassi, published in the volume of Fig. 6.3, pp. 61–73, Bologna, Biblioteca Comunale dell'Archiginnasio

OSSERVAZIONI
FISICO - MEDICHE
INTORNO ALLA
ELETTRICITA'
DEDICATE
ALL' ILLUSTRISSIMO, ED ECCELSO
SENATO
DI BOLOGNA
DA GIO: GIUSEPPE VERATTI
Pubblico Professore nella Università,
E nell' Accademia delle Scienze dell' Instituto
Accademico Benedettino.

IN BOLOGNA

Nella Stamperìa di Lelio dalla Volpe. 1748.
Con licenza de' Superiori.

Fig. 6.5 Frontespice of the book by Gio: Giuseppe Veratti, *Osservazioni fisico-mediche intorno all'elettricità*, Stamperia della Volpe, Bologna 1748, Bologna, Biblioteca Comunale dell'Archiginnasio

Fig. 6.6 A Leyden jar from the Cowper Collection, purchased for the institute in 1790, Bologna, Museo di Palazzo Poggi

Fig. 6.7 A Ramsden electrostatic machine from the Cowper Collection, Bologna, Museo di Palazzo Poggi

Fig. 6.8 Volta's pistol (eighteenth century), Bologna, Museo di Palazzo Poggi

References

Francesco Algarotti, *Il Newtonianismo per le dame, ovvero Dialogbi sopra la luce e i colori*, Naples [but Milan] 1737, p. III.

G. Angelelli, *Notizie dell'origine e progressi dell'Istituto delle Scienze di Bologna e sue accademie*, Stamperia dell'Istituto delle Scienze, Bologna 1780, p. 110.

Annarita Angelini (Editor), Anatomie Accademiche, vol III, *L'Istituto delle scienze e l'Accademia*, Il Mulino, Bologna 1993

Giulio Barsanti, *Spallanzani e le "resurrezioni". Rotiferi, tardigradi, angillule e altre "besticciuole"* in *La sfida della modernità. Atti del Convegno internazionale di. studi nel bicentenario della morte di Lazzaro Spallanzani,* edited by Walter Bernardi and Marta Stefani, Olschki, Firenze 2000, pp. 171-195, in part. pp. 179–188.

Giambattista Beccaria, *Elettricismo naturale e artificiale*, Stamperia di F.A. Campana, Turin 1753.

Giambattista Beccaria, *Elettricismo naturale. Lettere a Jacobo Bartolomeo Beccari*, Tipografia di Colle Ameno all'insegna dell'lride, Bologna 1758, pp. 29–30.

Bernardette Bensaude-Vincent and Christine Blondel (Editors) *Science and Spectacle in the European Enlightenment*, Ashgate, Aldershot, England - Burlington, USA 2008.

Maria Grazia Bergamini. *Interni d'accademia. Il sodalizio bolognese dei Vari. 1747–1763,* Mucchi, Modena 1996.

Gabriella Berti Logan, *The Desire to Contribute: An Eighteenth Century Italian Woman of Science*, «American Historical Review», 99.

Gabriella Berti Logan, *Women and the practice and teaching of Medicine in Bologna in the Eighteenth and early Nineteenth Centuries*, «Bulletin of the History of Medicine», 2003, 77, pp. 506–535: 517–534.

Paola Bertucci, *Viaggio nel paese delle meraviglie. Scienza e curiosità nell'Italia del Settecento*, Bollati Boringhieri, Torino 2005.

Fabio Bevilacqua and Lucio Fregonese (Editors), «Nuova Voltiana. Studies on Volta and his Times» II, 2000.

Marco Bresadola, *Luigi Galvani: devozione, scienza e rivoluzione*, Editrice Compositori, Bologna 2011, pp. 149–150 *passim.*

Marina Calore (Editor), *I padroni della villa. La famiglia Aldrovandi Marescotti nel Settecento*, La Tipografia Moderna, Bologna 1994

Marta Cavazza, Settecento inquieto. *Alle origini dell'Istituto delle Scienze di Bologna*, Il Mulino, Bologna 1990.

Marta Cavazza. *L'insegnamento, delle scienze sperimentali nell'Istituto delle scienze di Bologna*, «Alma Mater Studiorum», 1993, pp. 155–168 (English translation, *ibid.*, pp. 169–179).

Marta Cavazza, *Laura Bassi e il suo gabinetto di fisica sperimentale: realtà e mito*, «Nuncius, Annali di storia della scienza», X, 2, 1995a.

Marta Cavazza, *Orti botanici, teatri anatomici, musei e gabinetti scientifici,* in *Le università dell'Europa. Le scuole e i maestri. L'età moderna,* edited by G. P. Brizzi and J. Verger, A. Pizzi, Milano 1995b, pp. 67–89.

M. Cavazza, *L'Istituto delle scienze di Bologna negli ultimi decenni del Settecento,* in *La politica della scienza. Toscana e stati italiani nel tardo Settecento,* edited by Giulio Barsanti, Vieri Becagli. Renato Pasta, Olschki, Florence 1996, pp. 435–450.

Marta Cavazza, *La recezione della teoria halleriana dell'irritabilità nell'Accademia delle scienze di Bologna*, «Nuncius. Annali di storia della scienza», XII, 2, 1997a, pp. 359–377.

M. Cavazza, *"Dottrici"e lettrici nell'Università di Bologna nel Settecento,* «Annali di storia delle università italiane», I, 1997b, pp. 109–126, in part. pp. 113–118.

Marta Cavazza, Laura Bassi *«maestra» di Spallanzani, in Il cerchio della vita. Materiali di ricerca del Centro Studi Lazzaro Spallanzani di Scandiano sulla storia della Scienza del Settecento,* edited by Walter Bernardi and Paola Manzini, Olschki, Firenze 1999.

M. Cavazza, *Una donna nella Repubblica degli scienziati: Laura Bassi e i suoi colleghi*, in *Scienza a due voci*, a cura di Raffaella Simili, Olschki, Firenze 2005, pp. 61–85.

M. Cavazza, *Innovazione e compromesso. L'Istituto delle scienze e il sistema accademico bolognese del Settecento, in Storia di Bologna secoli XVII e XVIII. Bologna nell'età moderna. II, Cultura, istituzioni culturali, Chiesa e vita religiosa,* edited by Adriano Prosperi, Bononia University Press, Bologna 2008a, pp. 317–374.

Marta Cavazza, *Vis irritabilis e spiriti animali. Una disputa settecenteca sulle cause del moto muscolare,* in *Neuroscienze contrverse. Da Aristotele alla moderna scienza del linguaggio,* edited by Marco Piccolino, Bollati Boringhieri, Turin 2008b, pp. 49–74.

M. Cavazza, *Laura Bassi and Giuseppe Veratti: an electric couple during the Enlightenment,* «Contribution to Science», 5 (1), 2009a.

Marta Cavazza, *Between Modesty and Spectacle: Women and Science in Eighteenth-Century Italy,* in *Gender and Culture in Italy in the Age of the Grand Tour,* edited by Paula Findlen, Catherine Sama, Wendy Roworth, University of Stanford Press, Stanford 2009b, pp. 275–302.

Marta Cavazza, *Laura Bassi. Donne, genere e scienza nell'Italia del Settecento,* Editrice Bibliografica, Milano, 2020.

Beate Ceranski (Editor), *Il carteggio tra Giovanni Bianchi e Laura Bassi, 1733–1745,* «Nuncius», IX, 1994, l, pp. 207–231.

Beate Ceranski, *"Und sie fürchet sich vor niemandem" Über die Physikerin Laura Bassi (1711–1778),* Campus, Frankfurt-New York 1996.

Charles De Brosses, *Lettres d'Italie du Président De Brosses,* sous la direction de Frédéric d'Agay, Mercure de France, Paris 1986, 2 voll., l, pp. 267–268.

Pericle Di Pietro (Editor), *Edizione Nazionale delle opere di Lazzaro Spallanzani, Carteggi,* vol I. Mucchi, Modena 1984

Pericle Di Pietro (Editor), *Edizione Nazionale delle Opere di Lazzaro Spallanzani, Parte IV. Opere edite direttamente dall'Autore,* vol. I, Mucchi, Modena 1996.

Giorgio Dragoni, *Vicende dimenticate del mecenatismo bolognese dell'ultimo '700: l'acquisto della collezione di strumentazioni scientifiche di Lord Cowper,* Il Carrobbio, XI, 1985, pp. 68–85.

Giovanni Fantuzzi, *Elogio della dottoressa Laura Maria Caterina Bassi Veratti,* Stamperia S. Tommaso d'Aquino, Bologna 1778.

Giovanni Fantuzzi, *Notizie degli scrittori bolognesi,* Stamperia S. Tommaso d'Aquino, Bologna 1781–1790, 9 vol.

Paula Findlen, *Science as a career in Enlightenment Italy. The strategies of Laura Bassi,* «Isis», 84, 1993, pp. 441–469.

Paula Findlen, *The Scientist's Body: The Nature of a Woman Philosopher in Enlightenment Italy,* in *The Faces of Nature in Enlightenment Europe,* edited by Lorraine Daston, Gianna Pomata, BWV, Berlin 2003, pp. 234–236.

Miriam Focaccia (Editor), *Anna Morandi Manzolini. Una donna fra arte e scienza. Immagini, documenti, repertorio anatomico,* Olschki, Firenze 2008.

Roberto Gandini, *Fedele da Scandiano, intagliatore e tecnico-fisico di Lazzaro Spallanzani nelle Università di Pavia e di Modena,* Editrice Age, Reggio Emilia 1972.

Antonio Garelli (Editor), *Lettere inedite alla celebre Laura Bassi scritte da illustri italiani e stranieri,* Tip. Cenerelli, Bologna 1885.

John L. Heilbron, *Alle origini della fisica moderna. Il caso dell'elettrictà,* Italian translation, Il Mulino, Bologna 1982 (Berkeley-Los Angeles 1979).

Francesco Massardi (Editor), *Alessandro Volta, Edizione Nazionale delle Opere, Epistolario,* I vol., Zanichelli, Bologna 1949, p. 187.

Serafino Mazzetti, *Repertorio di tutti i professori antichi e moderni della famosa Università e del celebre Istituto delle Scienze di Bologna,* Tipografia di S. Tommaso d'Aquino, Bologna 1847.

Mario Medici, *Elogio di Luigi Galvani,* Tipografia governativa Alla Volpe, Bologna 1845.

Elio Melli (Editor), *Epistolario di Laura Bassi Veratti, in Studi e inediti per il primo centenario dell'Istituto Magistrale «Laura Bassi»,* Tip. STEB, Bologna 1960.

Elio Melli, *Laura Bassi Veratti: ridiscussioni e nuovi spunti,* in *Alma mater studiorum. La presenza femminile dal XVIII al XX secolo. Ricerche sul rapporto donne/cultura universitaria nell'Ateneo bolognese.* CLUEB, Bologna 1988, pp. 71–80.

Rebecca Messbarger, *The Lady Anatomist: The life and Work of Anna Morandi Manzolini*, The Chicago University Press, Chicago and London 2010.

Antonio Pace, *Benjamin Franklin and Italy*, The American Philosophical Society, Philadelphia 1958.

Giuliano Pancaldi, *Volta: Science and Culture in the Age of the Enlightenment*, Princeton University Press, Princeton-Oxford 2003.

Helena M. Pycior, Nancy G. Slack, Pnina G. Abir-Am (Editors), *Creative couples in the sciences*, Rutgers University Press, New Brunswiek, NJ 1996.

Moira R. Rogers, *Newtonianism for the Ladies and Other Uneducated Souls. The Popularisation of Science in Leipzig, 1687–1750*, Peter Lang, NewYork, 2003.

Gian Antonio Salandin, Maria Pancino, *Il "teatro" di filosofia sperimentle di Giovanni Poleni*, LINT, Trieste 1987.

Walter Tega (Editor), *Anatomie Accademiche*, vol I, *I Commentari dell'Accademia delle scienze di Bologna*, Il Mulino, Bologna 1986, and vol II, *L'enciclopedia scientifica dell'Accademia delle scienze di Bologna*, Il Mulino, Bologna 1987.

Giuseppe Venturoli, *Elogio del Sig. Gian Francesco Malfatti*, «Memorie di matematica e fisica della Società italiana delle scienze», XV, parte prima, 1811, pp. XXVI-XXXVI.

Giuseppe Veratti, *Osservazioni fisico-mediche intorno alla Elettricità*, Lelio dalla Volpe, Bologna 1748.

Giuseppe Veratti, *Osservazioni fatte in Bologna l'anno 1752 dei fenomeni elettrici nuovamente scoperti in America e confermati in Parigi*, L. Dalla Volpe, Bologna (1752).

Credits

Front Cover: Courtesy of "Alma Mater Studiorum"—Sistema Museale d'Ateneo—Museo di Palazzo Poggi, Bologna, ref. No. 0000374, 26/05/2020.
Back Cover: Biblioteca Comunale dell'Archginnasio, Bologna, ref. No. 95/IV-31, 8/02/2020.
Figures 1.4, 4.8, 4.9, 4.10, 4.11, 6.1, 6.2: Courtesy of Ministero per i Beni e le Attività Culturali—Archivio di Stato di Bologna.
Figures 1.5, 5.4, 6.3, 6.4, 6.5: Biblioteca Comunale dell'Archiginnasio, Bologna, ref. No. 95/IV-31, 8/02/2020.
Figures 1.1, 1.2, 2.1, 4,4, 4.5: © Alma Mater Studiorum, Università di Bologna—Biblioteca Universitaria di Bologna, refs. No. 0000174, 20/02/2020 and No. 0000207, 27/02/2020.
Figure 2.2: © The Fitzwilliam Museum, Cambridge.
Figures 4.1, 4.2, 4.3, 4.6, 4.7, 5.1, 6.6, 6.7, 6.8: Courtesy of "Alma Mater Studiorum"—Sistema Museale d'Ateneo—Museo di Palazzo Poggi, Università di Bologna, ref. No. 0000231, 9/03/2020.
Figures 3.3, 3.5, 3.8: Courtesy of Museo di Storia della Fisica, Università di Padova.

L. Cifarelli and R. Simili (eds.), *Laura Bassi–The World's First Woman Professor in Natural Philosophy*, Springer Biographies,
https://doi.org/10.1007/978-3-030-53962-7

Index of Names

A
Abbri, Ferdinando, 19, 131
Abir-Am, Pnina G., 132
Agnesi, Maria Gaetana, 12, 24, 82, 83, 95, 97, 106, 107, 110
Alberts, R. Chr, 19
Aldrovandi Marescotti, Filippo, 84, 85, 115, 116, 134, 135
Alembert Jean Baptiste d', 24, 26, 41
Algarotti, Francesco, 42, 71, 72, 81, 89, 96, 128, 132
Ampère, André-Marie, 66
Angelelli, Giuseppe, 130
Angelini, Annarita, 2, 5, 7, 11, 13, 26, 70, 120, 127, 130
Ardinghelli, Maria Angela, 97
Aristotle, 71
Arrighi, Gino, 106
Ashe, George, 13

B
Babini, Valeria Paola, 11, 106
Bacchi, M.C., 10, 14
Bacon, Francis, 22
Baiada, Enrica, 10, 11
Balbi, Paolo, 75, 84
Barbapiccola, Eleonora, 101
Barsanti, Giulio, 119, 120
Bassi, Giuseppe, 70, 125
Bazzani, Matteo, 71, 80, 98
Becagli, Vieri, 120
Beccaria, Giambattista, 12, 24, 27, 83, 119, 129–131
Beccari, Jacopo Bartolomeo, 26, 27, 88, 95, 98, 101, 121, 129
Belloni, Luigi, 98
Benassi, S., 21
Benedetto (Benedict) XIV, Pope, 7, 11, 12, 22–24, 26, 76–78, 86, 88, 96, 107, 116, 120, 125, 133
Bennet, Jim, 62
Bensaude Vincent, Bernardette, 131, 132
Bentivoglio Davia, Laura, 72, 95, 99–102
Beretta, Marco, 131
Bergamini, Maria Grazia, 121
Bernardi, Walter, 102, 119
Bernoulli, Daniel, 41
Bernoulli, Johann, 37, 39
Berti Logan, Gabriella, 69, 71, 84, 116
Bertucci, Paola, 75, 101, 126, 128, 132
Bevilacqua, Fabio, 131
Bianchi, Giovanni, 95, 97, 100–106, 108, 111, 123
Bianconi, G.G., 8
Bignon, Abbé, 21
Biot, Jean-Baptiste, 66
Black, Joseph, 84
Blondel, Christine, 132
Boerhaave, Hermann, 5, 9, 10, 14–21, 118
Bolletti, Giuseppe Gaetano, 10, 11, 78
Bònoli, Fabrizio, 10
Bortolotti, Ettore, 3
Boscovich, Ruggero (Roger), 24, 27, 84
Bose, Georg Mathias, 64
Bovier de Fontenelle, Bernard le, 5, 14, 21, 96
Boyer, Jean Baptiste, 104
Boyle, Robert, 10, 16, 18, 22, 51, 52, 82, 128
Braccesi, Alessandro, 10
Brackenridge, Bruce, 36
Bradley, James, 38

L. Cifarelli and R. Simili (eds.), *Laura Bassi–The World's First Woman Professor in Natural Philosophy*, Springer Biographies,
https://doi.org/10.1007/978-3-030-53962-7

Bresadola, Marco, 102, 130
Brizzi, Giampaolo, 121
Brosses, Charles de, 123
Buffon, George Louis, 24
Buonaccorsi, Giovanni Lorenzo, 85
Burnet, Thomas, 14

C
Calabro, Giovanni Bernardo, 97
Caldani, Marc'Antonio Leopoldo, 27, 83, 128, 129
Caldelli, M.L., 11
Calore, Marina, 116
Campeggi, Angiola, 116
Campi, Giuseppe, 131, 132
Candler Hayes, Julie, 95
Canterzani, Sebastiano, 26, 27, 82, 84, 85, 118, 127, 128
Cappelletti, Vincenzo, 98
Casini, Paolo, 42
Cassini, Gian Domenico, 4, 98, 106
Cavalieri, Bonaventura, 98
Cavazza, Marta, 12, 35, 42, 69, 70, 73–76, 81, 83, 85, 102, 115–121, 123, 125, 128, 129, 131
Ceranski, Beate, 35, 69, 74, 102–106, 116, 123, 127, 128
Cesi, Federico, 99
Changuion, Franç, 19
Christina of Sweden, queen, 11
Clairaut, Alexis, 24, 41, 84
Clemente XI, Pope, 7
Clemente XIV, Pope, 99
Cohen, I. Bernard, 38
Collina, Father Abundio, 69
Colonna, Fabio, 97
Corazzi, Ercole, 2, 11
Cosmopolita, Simone, 107. *See also* Bianchi, Giovanni
Cowper, George, 116, 139
Cremante, R., 5
Crivelli, Giovanni, 103

D
d'Agay Frédéric, 123
Dalla Bella, Giovanni Antonio, 62
Dalle Donne, Maria, 86, 116
Daston, Lorraine, 74
Davia, Giovanni Antonio, 97
De Carolis, Stefano, 99, 100, 108
De Clercq, Peter, 53, 54
De Hondt, P., 19
del Buono, Girolamo, 106
De Limières, Henri Philippe, 20, 21
Del Negro, Piero, 62
Derham, William, 14
Desaguliers, John Theophilus, 53, 55
Descartes, René, 43, 71, 97, 101, 102
De Zan, Mauro, 70, 72, 80
Diderot, Denis, 19
Di Martino, Pietro, 73
Di Pietro, Pericle, 119, 120, 126, 130
Di Trocchio, Federico, 98
Dollond, John, 84
Donati, Giancarlo, 99
Dragoni, Giorgio, 116
du Boccage, Anne-Marie, 81
Du Châtelet, Emile, 24, 80, 81
Dufay, de Cisternay Charles François, 55, 64

E
Einmart, Giorgio Cristoforo, 10
Ekman, Martin, 41
Emiliani, Andrea, 10
Ercolani, Ratta Elisabetta, 72
Euler, Leonard, 39–41, 96

F
Fabi, Angelo, 99, 107
Fantuzzi, Giovanni, 71, 85, 117, 122, 125
Faraday, Michael, 66
Fedele da Scandiano, frair, 118, 119
Ferdinando II, grand duke, 51
Findlen, Paula, 12, 35, 69, 72, 74, 82, 86, 96, 97, 101, 102, 104, 107, 116, 129
Focaccia, Miriam, 102, 107, 122
Fontana, Felice, 83, 120, 125, 128, 130, 131
Formey, J. H. Samuel, 24
Foucault, Jean Bernard Léon, 66
Franceschini, Marcantonio, 20
Franklin, Benjamin, 84, 117, 129–131
Fregonese, Lucio, 131
Fresnel, Augustin Jean, 66
Frisi, Antonio Francesco, 107
Frisi, Paolo, 24, 27

G
Galeazzi, Domenico Gusmano, 71, 74–76, 78–80, 82, 84, 90, 95, 106, 108, 121, 122
Galilei, Galileo, 51, 52, 54
Galli, Giovanni Antonio, 121, 122

Galvani, Luigi, 12, 24, 26, 27, 83, 108, 121, 122, 130
Gandini, Roberto, 119
Garelli, Antonio, 119, 123, 125, 126, 130–133
Gassend, Pierre, 97
Gherardi, R., 9
Ghiselli, Antonio Francesco, 5
Ghisilieri, Filippo Carlo, 121
Giacomelli, Alfeo, 96
Giambelli, Carlo, 99
Giovenardi, Giampaolo, 99, 100
Gomez Ortega, Casimiro, 125
Gosse, P., 19
Gozza, P., 11
Grandjean de Fouchy, Jean-Paul, 62, 73
Gray, Stephen, 64
Greenberg, John L., 40
Gua de Malves, Jean Paul de, 35, 36
Guglielmini, Domenico, 14, 21, 26, 39, 98
Guidalotti, Gioseffo, 6

H
Hales, Stephen, 84
Haller, Albert von, 64, 95, 128–130
Halley, Edmond, 13
Hartley, John, 9
Hauksbee, Francis, 53, 64
Heilbron, John L., 55, 64, 121, 130
Hobbes, Thomas, 51
Hodgson, James, 53
Holmes, Fréderic, 131
Hooke, Robert, 42, 43
Hortega, José, 119, 125
Huygens, Christiaan, 9, 36, 38, 39, 42

I
Ingenhousz, Jan, 131

J
Jones, William, 37

K
Keill, John, 52, 53
Kuhn, Thomas, 42

L
Laghi, Tommaso, 108, 121, 122
Lagrange, Joseph Louis, 41
Lambertini, Prospero, 22, 23, 70, 71, 76, 88, 96. *See also* Bendetto XIV, pope
Lami, Giovanni, 106
Landriani, Marsilio, 131
Laplace, Pierre Simon, 40, 41
Laurenti, Marc'Antonio, 80
Lavoisier, Antoine Laurent, 84
Le Comte, Marguerite, 81
Leibniz, Gottfried Wilhelm, 9, 18, 36, 37
Lelli, Ercole, 76, 95
Leprotti, Antonio, 72, 76, 79, 98, 102, 105
Lindeboom, G. A., 19
Longhena, M., 21
Louis XIV, King of France, 51
Lyonnet, Pieter, 118

M
Maffioli, Cesare S., 37, 83
Magnan Campanacci, Ilario, 72
Magnani, Paolo, Cardinal, 23, 24
Malagida, Seniore, 118
Malfatti, Gian francesco, 125
Malpighi, Marcello, 2, 4, 14, 21, 98, 128
Mamiani, Maurizio, 43
Manfredi, Eustachio, 8, 26, 27, 36, 38, 69, 73, 80, 90, 98, 123
Manfredi, Gabriele, 37, 71, 72, 103
Manfredi, Maddelena, 72
Manfredi, Teresa, 72
Manzini, Paola, 119, 125, 131
Manzolini, Giovanni, 24, 95, 108, 122
Marchesini, Ferdinando, 122
Marini, Tommaso, 101
Marino, I., 11
Mariotte, Edme, 39, 128
Marsili, Anton Felice, 4, 6
Marsili, Luigi Ferdinando, 1–24, 26–29, 31, 32, 70, 96, 133
Masetti Zannini, Gian Ludovico, 102, 105
Masi, Ernesto, 77, 80
Massardi, Francesco, 132
Maupertuis, Pierre Louis, 24, 26, 96
Mazzetti, Serafino, 115, 122
Mazzotti, Massimo, 71, 80, 82, 107
Mc Lellan, James E., 73
Mead, Richard, 13
Medici Leopoldo de', Prince, 51
Medici, Mario, 122
Medici, Michele, 71
Melli, Elio, 75, 77–80, 116, 123–125, 127, 131, 133
Messbarger, Rebecca, 107, 108, 122

Micheli, Lorenzo, 118
Minuz, E., 11
Molinelli, Pier Paolo, 26, 108, 129
Montanari, Antonio, 97, 99
Montanari, Geminiano, 2, 21, 98
Monti, Gaetano, 26
Monti, Giuseppe, 26, 98, 105
Morandi Manzolini, Anna, 96, 97, 107–110, 122
More, Henry, 43
Morgagni, Giovan Battista, 6, 20, 21, 26, 95, 98
Morton, Alan Q., 53
Mozzi, Bartolomeo, 128
Müller, Johann, 20
Musschenbroek, Pieter van, 9, 14, 53, 55, 62

N

Naudé, Gabriel, 9
Neri, L., 11
Newton, Isaac, 9, 10, 13, 14, 18, 22, 24–27, 35–46, 52–55, 62, 63, 71, 72, 77, 84, 96, 97, 102, 117, 118, 123
Nollet, Jean Antoine, 49, 50, 55–62, 64–67, 75, 83, 117–119, 131

O

Odoardi, Ignazio, 126
Oldenburg, Henry, 13
Ongaro, Giuseppe, 98
Ortes, Gian Maria, 125

P

Pace, Antonio, 129
Pallas, Athena, 5, 15
Pancaldi, Giuliano, 93, 131
Pancino, Maria, 121
Pasini, Walter, 100
Pasta, Renato, 120
Peggi, Pier Francesco, 80
Pepe, Luigi, 37
Peruzzi, Giulio, 64
Piccolino, Marco, 102
Pignatelli, Faustina, 72, 73, 97
Pilastri Giambattista, count, 100, 108, 109
Pio VI, Pope, 99
Piro, Anna, 100
Poggi Giovanni, Cardinal, 2
Poleni, Giovanni, 54, 57, 61, 62, 121
Poletti, Luigi, 116
Polinière, Pierre, 52, 55
Pomata, Gianni, 74
Powers, John C., 19
Pozzi, Giuseppe, 106
Priestley, Joseph, 84
Prosperi, Adriano, 120
Pycior, Helena M., 132

R

Ranuzzi, Girolamo, 122
Ray, John, 13
Réaumur, René Antoine Ferchault de, 55
Riccati, Vincenzo, 26, 36
Rizzetti, Giovanni, 42
Rodati, Luigi, 116
Rogers, Moira R., 133
Rondelli, Geminiano, 98
Rosen Richard L., 70, 73, 74
Rossi, Antonio, 20
Rossi, P., 18, 19

S

Saladini, Girolamo, 36
Salandin, Gian Antonio, 62, 121
Sama, Catherine, 101, 123
Sauberf (alias Giobert Giovanni Antonio), 118
Scarselli, Flaminio, 77–80, 124, 125
Scheuchzer, J.J., 19
Seligardi, Raffaella, 131
Serra, Angelo, 108
Serra, Carlo Antonio, 108
Serra, Giuseppe, 108
's Gravesande, Willem Jacob, 53–56, 62
Sherard, William, 13
Simili, Alessandro, 95, 111
Simili, Raffaella, 12, 97, 101, 102
Slack, Nancy G., 132
Sloane, Hans, 13
Smith, George, 39
Spallanzani, Lazzaro, 12, 83, 85, 118–120, 125, 126, 130, 131
Stancari, Vittorio Francesco, 8
Stefani, Marta, 119
Stoye, J., 5, 11, 13, 14, 20, 70
Sydenham, Thomas, 13
Symmer, Robert, 129, 130

T

Tacconi, Gaetano, 70–72, 102, 106
Taglianini, S., 11
Talas, Sofia, 62, 64

Talman, M., 10
Tegarelli, Antonio, 100
Tega, Walter, 5, 11, 12, 19, 23, 26, 27, 70, 76, 99, 120, 129
Tiberius, Emperor, 20
Toldi, Pietro, 116
Tonini, Carlo, 97
Torricelli, Evangelista, 39
Trionfetti, Lelio, 2, 6, 8, 98
Trucco Emanuela, 43
Truesdell, Clifford, 40
Turnbull, Herbert W., 42

U

Urbinati, Nadia, 11, 76, 129
Uytwerf, Herm, 19

V

Vallisneri, Antonio, 95
Valsalva, Antonio, 26, 98
Van Swieten, Gerard, 95
Venturoli, Giuseppe, 125
Veratti, Francesco, 118
Veratti, Giuseppe, 74, 101, 115, 122, 128, 133, 138
Veratti, Paolo, 86, 115
Verger, Jacques, 121
Verzaglia, Giuseppe, 37, 128
Vittuari, Francesco, 118
Volder, Byrchard de, 52
Volta, Alessandro, 83, 117–119, 131, 132, 139
Voltaire (Arouet, F.-M.), 24, 26, 62, 63, 77, 80, 95

W

Wassyng Roworth, Wendy, 101, 123
Watelet, Claude-Henri, 81
Wess, Jane A., 53
Whiston, William, 52
Woodward, John, 14

Z

Zambeccari, Giacomo, 12
Zanotti, Eustachio, 26, 71
Zanotti, Francesco Maria, 25–27, 42, 69, 70, 86, 89, 101
Zeus, 5
Zinsser, Judith P., 95

Printed in the United States
by Baker & Taylor Publisher Services

Printed in the United States
by Baker & Taylor Publisher Services